Benchmark Papers in Systematic and Evolutionary Biology Series

Editor: Carl Jay Bajema — Grand Valley State College

EVOLUTION BY SEXUAL SELECTION THEORY
Prior to 1900

Edited by

CARL JAY BAJEMA
Grand Valley State College

A Hutchinson Ross Benchmark® Book

 VAN NOSTRAND REINHOLD COMPANY

Copyright © 1984 by **Van Nostrand Reinhold Company Inc.**
Benchmark Papers in Systematic and Evolutionary Biology, Volume 6
Library of Congress Catalog Card Number: 84-3538
ISBN: 0-442-21181-3

Manufactured in the United States of America.

Published by Van Nostrand Reinhold Company Inc.
135 West 50th Street
New York, New York 10020

Van Nostrand Reinhold Company Limited
Molly Millars Lane
Wokingham, Berkshire RG11 2PY, England

Van Nostrand Reinhold
480 Latrobe Street
Melbourne, Victoria 3000, Australia

Macmillan of Canada
Division of Gage Publishing Limited
164 Commander Boulevard
Agincourt, Ontario MIS 3C7, Canada

15   14   13   12   11   10   9   8   7   6   5   4   3   2   1

Library of Congress Cataloging in Publication Data
Main entry under title:
Evolution by sexual selection theory.
   (Benchmark papers in systematic and evolutionary biology; v. 6)
   "A Hutchinson Ross Benchmark book."
   Includes index.
   1. Evolution—Addresses, essays, lectures.   2. Sexual selection in animals—
Addresses, essays, lectures.   I. Bajema, Carl Jay.   II. Series.
QH371.E923   1984   575.01'62   84-3538
ISBN 0-442-21181-3

CONTENTS

Contents

PART IV: CONTROVERSY OVER *THE DESCENT OF MAN AND SELECTION IN RELATION TO SEX,* 1871-1882

Contents

SERIES EDITOR'S FOREWORD

The volumes of the Benchmark Papers in Systematic and Evolutionary Biology series reprint classic scientific papers on the evolution and systematics of organisms. These volumes do more than just provide scholars with facsimile reproductions or English translations of classic papers on a particular topic. The interpretative commentaries and extensive bibliographies prepared by each volume editor provide busy scholars with a review of the primary and secondary literature of the field from a historical perspective and a summary of the current state of the art.

The theories in which the environment plays a crucial role in giving direction to evolution can be divided into three categories: *Geoffroyism* — the production of heritable adaptive variations by the environment acting directly on the organism; *Lamarckism* — the production of heritable adaptive variations as the result of the increasing use or disuse of organs by an organism as it experiences new needs or the alteration of existing needs due to changes in its environment; and *neo-Darwinism* — the production of both harmful and beneficial heritable variations in organisms by causes unrelated to the specific adaptive problems the organisms are encountering followed by the selective reproduction of heritable variations brought about by the ecological interactions each organism has with its environment. Both Lamarckism and Geoffroyism have been scientifically falsified by modern scientific knowledge concerning the production of gene mutations and their transmission to offspring. Only the Darwinian theory of adaptive evolution by natural, sexual, and artificial selection has been demonstrated to be consistent with the modern scientific understanding of genetics.

In *Evolution by Sexual Selection Theory: Prior to 1900*, the history of Charles Darwin's theory of sexual selection is traced. The book begins with the speculations of Thomas Hobbes and William Harvey in the seventeenth century, follows through the construction and defense of Darwin's theory, and concludes with the two decades immediately following Charles Darwin's death in 1882 when attack on sexual selection theory was increasing. This volume, the sixth in the series, is the third of eight volumes on selection theory being edited by Carl Bajema. Volumes on *Artificial Selection and the Development of Evolutionary Theory* and *Natural Selection Theory from the Speculations of the Greeks to the Quantitative Measurements of the Biome-*

tricians have already been published. One of the additional volumes on selection theory to be published will be devoted to sexual selection theory in the twentieth century and the development of sociobiology. Benchmark Papers in Systematic and Evolutionary Biology volumes on biogeography, comparative anatomy, comparative embryology, interspecific competition, and other related topics are also being planned.

CARL JAY BAJEMA

PREFACE

Charles Darwin's theory of adaptive evolution by natural, sexual, and artificial selection is the most important and yet the most widely misunderstood principle in biology. The volumes on selection theory that I am editing are designed to help students and scholars gain a better understanding of the history of the controversy surrounding the relative importance of selection as an ecological process generating adaptive evolution and the current state of our scientific understanding of the role that selection plays in giving direction to evolution.

The theory that adaptive evolution occurs as the result of selection operating on heritable variations was proposed by Charles Robert Darwin and Alfred Russel Wallace in 1858. *On the Origin of Species by Means of Natural Selection, or the Preservation of Favoured Races in the Struggle for Life,* which Charles Darwin published the following year, triggered a scientific revolution with respect to our understanding of the origin and continuing evolution of species.

Artificial Selection and the Development of Evolutionary Theory (vol. 4 in the Benchmark Papers in Systematic and Evolutionary Biology series) describes how Charles Darwin, by reading the literature on the breeding of domesticated plants and animals, was led to the idea that selection could bring about evolution. It was only after Darwin had become familiar with selection as a process that could cause "descent with modification" (evolution) in domesticated species that he realized how powerful selection operating in nature was. Darwin became convinced that selection was powerful enough to bring about evolution when he read *An Essay on the Principle of Population* by Malthus. Darwin concluded that the power of selection to bring about evolution in nature resided in competition, the "struggle for existence" produced by the overproduction of offspring. The volume on *natural selection theory* traces the development of the theory of natural selection from the speculations of the ancient Greeks through the expansion of the theory to explain evolution by Charles Darwin and Alfred Wallace to the last decade of the nineteenth century when the biometricians developed statistical methods, which they used to directly measure the direction and intensity of ongoing selection in nature.

Evolution by Sexual Selection Theory: Prior to 1900 traces the development of Charles Darwin's theory of evolution by sexual selection from the speculations of Thomas Hobbes and William Harvey in the seventeenth century through the periods when Charles Darwin constructed his theory (1837-1842) and publicly advocated and defended his theory (1858-1882) to the two decades following his death in 1882 when attack on sexual selection theory increased. It is impossible to publish all the important papers on sexual selection theory in one manageable volume. A companion volume on sexual selection theory in the twentieth century and the rise of sociobiology is in preparation. Volumes on the theory and measurement of natural selection in the twentieth century, domestication of plants and animals by selection, human evolution by selection, and cultural evolution by selection are also in preparation.

I wish to thank the people who provided assistance while I selected papers for inclusion in this volume and wrote the editor's commentaries on the history of natural selection theory: Peter Gautrey, University Library, Cambridge; David Kohn, Harvard University; Will Provine, Cornell University; Michael Ruse, University of Guelph; Eugenie Scott, University of California; Mary Jane West-Eberhard, Smithsonian Tropical Research Institute; E. O. Wilson, Harvard University; William Wimsatt, University of Chicago; and the librarians at Grand Valley State College and elsewhere who located and obtained copies of numerous scholarly publications for me through interlibrary loans. I also wish to thank the faculty and staff at the University of Colorado who organized the Third Regional Conference on the History and Philosophy of Science in April 1979, where I first presented a paper reconstructing the history of Charles Darwin's theory of sexual selection.

CARL JAY BAJEMA

CONTENTS BY AUTHOR

EVOLUTION BY SEXUAL
SELECTION THEORY
Prior to 1900

Part I

SEXUAL SELECTION THEORY
BEFORE CHARLES DARWIN

Editor's Comments
on Papers 1 Through 10

9   SEBRIGHT
Excerpt from *The Art of Improving the Breeds of Domestic
Animals. In a Letter Addressed to the Right Hon. Sir
Joseph Banks, K.B.*

10   PRICHARD
Excerpt from *On Diversities of Form*

The theory that contemporary organisms are the products of
"descent with modification" by natural selection, sexual selection,
and artificial selection (domesticated organisms only) was proposed
by Charles Darwin in 1858 and 1859. (For volumes on artificial selection
and natural selection, see Bajema, 1982, 1983.) The theory of sexual
selection was constructed by Charles Darwin in an attempt to explain
the evolution and continuing existence of intrasexual and intersexual
conflict among members of the same species with respect to
reproduction.

Many ancient writers contended that there was an idyllic time in
the past—a golden age—when all organisms lived in peace with one
another. The explanation for the demise of the golden age that came
to dominate Western thought was recorded in the Book of Moses
called *Genesis*. Moses contended that a demonic spirit initiated
human greed (sin), which destroyed this harmonious world, and that
human sin is responsible for the conflict that we observe among
individual organisms, particularly among human beings (Passmore,
1974, p. 7).

The scientific revolution that began in the sixteenth and seventeenth centuries involved changes in the types of causes some
scholars used to construct hypotheses explaining events and in the
way that these scholars tested their hypotheses. This scientific approach
to the construction and testing of hypotheses concerning the natural
causes and natural consequences of events was employed by a few
scholars to investigate the possible natural causes and consequences
of competition among members of the same species.

Thomas Hobbes (1588-1679) was the first philosopher to consistently
use the individual rather than the group as the basis for constructing
theories about political life. He contended that the animate world was
in constant struggle rather than being in a harmonious balance of
nature. In the *Leviathan* (1651, excerpts are presented as Papers 1A
and 1B), Hobbes argued that all men pursue their personal desires

(gain, security, and reputation), but since the resources men want are scarce, not all desires can be gratified (Ophuls, 1973, p. 50). The result is "war of every man against every man" (p. 62 in Paper 1A) among humans in a state of nature (the absence of any authorities or rules requiring men to keep the peace). Hobbes concluded that human passions make good intentions and voluntary compliance totally unreliable (Ophuls, 1973, p. 218; 1977, pp. 148-149).

A much more general statement of sexual selection theory operating in the human species and one that incorporates the Hobbesian argument that men "in a state of nature" pursue their own personal desires at the expense of other human beings has been published by Hardin (1968, 1974, 1977). He contends that human beings have been selected to behave in their own individual self-interests not only in the Hobbesian "state of nature" but also in every society, regardless of what social arrangements exist in that society. Hardin further contends that individual behavior that is harmful to society but beneficial to the individual and his family will be selected whenever political systems for distributing the benefits and costs of using resources do not couple the costs—the "responsibilities" or "obligations"—of performing an act such as pollution or reproduction with the benefits—the "rights" or "rewards"—of performing the act. Selection will favor individuals who behave selfishly and gain individual benefits (lower economic costs, more children) at the expense of the general social welfare. Appeals to conscience ("jawbone" responsibility) for voluntary restraint merely generate selection favoring those individuals who are unable or unwilling to limit their childbearing to the level that is in the best interest of society as a whole.

In 1651 the English physician and physiologist William Harvey (1578-1657) published *On the Generation of Animals,* which contains his observations on sexual competition (excerpts reprinted as Papers 2A, 2B, and 2C). It is interesting to note that Harvey identified the two ways in which Charles Darwin was to contend two centuries later that sexual selection operated—the male's power to charm his mates, and the male's power in combat with other males:

> Our common cock . . . is distinguished by his spurs, and ornamented with his comb and beautiful feathers, by which he charms his mates to the rites of Venus, and is furnished for the combat with other males, the subject of dispute being no empty or vainglorious matter, but the perpetuation of the stock in this line or in that; as if nature had intended that he who could best defend himself and his, should be preferred to others for the continuance of the kind. (pp. 425-426 in Paper 2C)

The explanation Harvey mentioned for the existence of such animal ornamentation, weaponry, and behavior is a species-benefit argument.

One can only speculate how the history of sexual selection theory might have been changed if William Harvey had lived long enough to write and publish his projected treatise on the *Loves, Lusts, and Sexual Acts of Animals.*

The English merchant and intellectual father of the science of demography, John Graunt (1620–1674), championed the idea that the development and use of population statistics would improve government. Graunt studied the death records that had been kept by the parishes of London since 1532 and which included age and cause of death, publishing his analyses of the causes of death and their relative *impact and constancy over time in 1662 in Natural and Political Observations Mentioned in a Following Index, and Made upon the Bills of Mortality (Graunt, 1899).* Graunt observed that although the male birthrate was higher than the female, it was offset by a higher mortality rate for males producing a population that was divided almost evenly by sex. He concluded that *"Christian Religion,* prohibiting *Polygamy,* is more agreeable to the *Law of Nature,* that is, the *Law of God,* than *Mahumetism* and others that allow it" (see p. 374 of Paper 3).

The English physician John Arbuthnot (1667–1735) employed *teleology*—the doctrine of final causes—to contend that a providential deity produced the observed slight surplus of male births over female births in human beings (Paper 4). He used data on religious christenings in London to construct a series of eighty-two annual ratios of male to female births from the years 1629 to 1710 inclusive and found that more males than females were born in each of the eighty-two years he studied. Arbuthnot concluded that the ratio of male to female births was not due to chance involving equal probability since the probability distribution for eighty-two years would have been $(\frac{1}{2} + \frac{1}{2})^{82}$, making it virtually impossible to have more births on one side of the binomial mean in all eighty-two years. The statistician Karl Pearson (1978, p. 132) observed that it did not seem to have occurred to Arbuthnot "that if the binomial distribution were $(p + q)^{82}$, his data might form also a chance distribution just as much as if $\frac{1}{2} + \frac{1}{2}$ were appropriate." Arbuthnot's appeal to final causes (teleology) as the cause of the slight surplus of male over female births has been summarized by Pearson (1978, p. 132):

> The greater supply of males he considers a wise economy of Nature, as the males are more subject to accidents and diseases, having to seek their food with danger. Therefore provident Nature to repair the loss brings forth more males. The near equality of the sexes is designed that every male may have a female of the same country and of suitable age. The persistency of male superiority of births shows that it is art and not chance that governs. Like John

Graunt, Arbuthnot uses the argument that Polygamy is contrary to the Law of Nature and Justice.

Charles Darwin (1871, 1874) considered whether "the tendency to produce the two sexes in equal numbers was advantageous to the species" and would be favored by selection. He concluded that the "whole problem is so intricate that it is safer to leave its solution for the future" (Darwin, 1874, pp. 259–260). The modern scientific understanding of how selection influences the sex ratio of a species is based on the theoretical analyses published by Fisher in 1930 (see also Weiss and Ballanoff, 1975).

The French philosopher and political theorist Jean-Jacques Rousseau (1712–1778) proposed the theory that ancestral man was naturally good in his *Discourse Upon the Origin and Foundation of the Inequality Among Mankind* (1755, excerpts from a 1761 English translation appear as Paper 5). Ancestral man, according to Rousseau, was naturally good because his desires were greatly limited as the result of ancestral man's solitary lifestyle and stupidity. Natural man, "deprived of every kind of enlightenment [lumieres]," was stupid. Consequently he was capable of knowing nothing but the simplest desires. All those desires that go beyond man's physical needs are "relative" desires that arise from a concern with the opinions or actions of other men. Since natural man lived a solitary life, he had no relative desires because these desires could arise only in a man who had social relations with other men (Plattner, 1979, p. 74).

Rousseau argued that inequality among human beings was the result of two causes—natural (physical) causes, and political (moral) causes:

> I conceive two Species of Inequality among Men; one which I can Natural, or Physical Inequality, because it is established by Nature, and consists in the difference of Age, Health, bodily Strength, and the Qualities of the Mind, or of the Soul; the other which may be termed Moral, or Political Inequality, because it depends on a kind of Convention, and is established, or at least authorized by the common consent of Mankind. This Species of Inequality consists in the different Privileges, which some Men enjoy, to the Prejudice of others, such as that of being richer, more honoured, more powerful, and even that of exacting obedience from them. (Rousseau, 1761, pp. 6–7)

Rousseau tried to prove that "the Inequality which may subsist between Man and Man in a State of Nature, is almost imperceivable, and that it has very little Influence" (p. 92 of Paper 5). He went on to argue that man's past natural goodness was rendered wicked as the result of the processes that made ancestral man sociable. Rousseau denied the belief that primitive man (as opposed to contemporary savage man) possessed the same sexual passions as social man, passions that would have forced Rousseau to accept the Hobbesian view that man's passions in a state of nature produced evil—"a war of all against all." Rousseau argued that man in a state of nature was incapable of making the observations and comparisons that were a prerequisite to developing the evil sexual passions characteristic of social man. In contrast to Hobbes who argued that human justice is not possible in a state of nature because contracts can be broken without sanction, Rousseau contended that ancestral human beings who lived in a state of nature lacked the intelligence to make contracts (Rousseau, 1762; Masters, 1968, p. 136). He also contended that no conclusions about the sexual passions of man in a state of nature could be drawn from other species, "from the Battles of the Males, who in all Seasons cover our Poultry Yards with Blood, and in Spring particularly cause our Forests to ring again with the noise they make in disputing their Females . . ." (pp. 83–84 in Paper 5). Rousseau's ideas concerning man in a state of nature have recently been reappraised by Masters (1978) and Wokler (1978). By setting social man in opposition to biological man, and by defending the proposition that ancestral human beings were naturally good, Rousseau led many scholars to neglect biological dimensions of human nature and search for the causes of human conflict solely in the structure and the functioning of human social institutions.

The purely philosophical quest to understand the nature of man undertaken by Hobbes and Rousseau stands in contrast to the Judeo-Christian theological appeal to divine revelation—the Genesis account of man's creation by God and Adam's original sin—to explain the origin of all evil (human conflict).

The theory that human ideas of beauty have affected marriage patterns so greatly that they have produced individual differences within and between human races was developed during the seventeenth, eighteenth, and nineteenth centuries. In 1686 the French traveler J. Chardin (1643–1713) contended that Persian ideas of beauty had affected the physical characteristics of the Persians by greatly influencing their marriage patterns (Paper 6).

In 1787 Samuel S. Smith (1750–1819), an American Presbyterian minister and president of Princeton College, published his *Essay on the Causes of the Variety of Complexion and Figure in the Human Species* in which he contended that the physical variety among the peoples of the world was due to natural causes—"effects of climate" and "the state of society," which included "diet, clothing, lodging, manners, habits, face of the country, objects of science, religion, interests, passions, and ideas of all kinds" (pp. 62–63 in Paper 7). He argued that beauty was an idea that influences "connexions in marriage" and produces differences in physical characteristics between different social classes as well as between different races. Smith was impressed with this "power which men possess over themselves of producing great changes in the human form, according to any common standard of beauty which they may have adopted" (p. 63 in Paper 7).

The theory that human ideas of beauty have affected mate selection and brought about evolutionary change among humans was also championed by the British physician-anthropologists James Prichard in 1813 (excerpted in Paper 10) and William Lawrence (1819). Stocking (1973, p. lvi), the historian of anthropology, has pointed out that Prichard's "notion of sexual selection was based on the assumption that a single (and of course European) standard of beauty had been 'implanted by Providence in our nature' to serve as a 'constant principle of improvement' " (p. 41 in Paper 10). Prichard did conclude that "it is probable that the natural idea of the beautiful in the human person has been more or less distorted in almost every nation" (p. 44 in Paper 10) and that the effects of such different national standards must play a very important role in the initial production and/or subsequent widening of human physical differences among nations (p. 45 in Paper 10). In 1839 Charles Darwin read a summary of Prichard's and Lawrence's ideas in *Intermarriage* (Walker, 1838), which may have helped him decide to divide selection operating in nature into natural selection and sexual selection.

Erasmus Darwin (1731–1802), Charles Darwin's grandfather, proposed a theory of evolution based on the inheritance of acquired characteristics in his book *Zoonomia* (1794–1796) several years before the great French biologist Jean Lamarck proposed his ideas on the same subject. Erasmus Darwin devoted two paragraphs to examples of animal "lust . . . the desire of the exclusive possession of the females" and the weapons males have acquired to combat each other for this purpose (p. 503 in Paper 8). Charles Darwin read his grandfather's *Zoonomia* while a student and also in 1837 when he first began recording ideas on evolution in his notebooks on the transmutation of species. Erasmus Darwin's explanation for sexual combat was both teleological in nature and based on the inheritance of characters acquired by exertion to satisfy desire. His theories about sexual combat were not the source of the grandson's theory of sexual selection, but they may have been one of the stimuli that led Charles Darwin to construct an alternative theory (Ghiselin, 1976, p. 127).

Intraspecific competition among humans was discussed by Thomas Robert Malthus (1798, 1803, 1826) in his classical *Essay on the Principle of Population.* . . . Malthus limited his discussion of intraspecific competition primarily to intergroup competition among

8

primitive tribes (Bowler, 1976, p. 636) as opposed to conflict within and among contemporary capitalist European societies. While Malthus emphasized struggle as opposed to cooperation in society, it was competition that operated for the benefit of all rather than a struggle for existence in which the weak suffered and the strong triumphed (Bowler, 1976, p. 644). He did not stress the importance of intraspecific struggle at the individual level that was the key ecological interaction that Darwin used to construct his theory of sexual selection.

Examples of all three forms of selection—artificial, natural, and sexual—that Darwin was to later divide selection into, were published by Sebright (1767–1846) in his 1809 pamphlet on *The Art of Improving the Breeds of Domestic Animals . . .* (excerpt in Paper 9). Darwin read Sebright's pamphlet several months before he read Malthus's *Essay on the Principle of Population* in September 1838 and was finally able to construct his theory of adaptive evolution by selection. The intellectual pathway that Charles Darwin took in constructing his theory that adaptations were the outcome of the operation of both sexual selection and natural selection in nature is discussed in Part II.

REFERENCES

Bajema, C. J., ed., 1982, *Artifical Selection and the Development of Evolutionary Theory,* Benchmark Papers in Systematic and Evolutionary Biology, vol. 4, Hutchinson Ross, Stroudsburg, Pa., 361p.

Bajema, C. J., ed., 1983, *Natural Selection Theory: From the Speculations of the Greeks to the Quantitative Measurements of the Biometricians,* Benchmark Papers in Systematic and Evolutionary Biology, vol 5. Hutchinson Ross, Stroudsburg, Pa., 384p.

Bowler, P. J., 1965, Malthus, Darwin, and the Concept of Struggle, J. Hist. Ideas **37:**631–650.

Darwin, C. R., 1858, Extract from an Unpublished Work on Species, by C. Darwin, Esq., Consisting of a Portion of a Chapter Entitled, "On the Variation of Organic Beings in a State of Nature; on the Natural Means of Selection; on the Comparison of Domestic Races and True Species," *Linnean Soc. London (Zool.) J. Proc.* **3:**46–50.

Darwin, C. R., 1859, *On the Origin of Species by Means of Natural Selection or the Preservation of Favoured Races in the Struggle for Life,* 1st ed., J. Murray, London, 502p. (facsimile edition by Harvard University Press, 1966).

Darwin, C. R., 1871, *The Descent of Man and Selection in Relation to Sex,* 1st ed., 2 vols., J. Murray, London, 423p., 475p.

Darwin, C. R., 1874, *The Descent of Man and Selection in Relation to Sex,* 2nd ed., J. Murray, London, 688p.

Fisher, R. A., 1930, *The Genetical Theory of Natural Selection,* 1st ed., Clarendon Press, Oxford, England, 272p.

Fisher, R. A., 1958, *The Genetical Theory of Natural Selection,* 2nd ed., Dover, New York, 291p.

Ghiselin, M., 1976, Two Darwins: History Versus Criticism, *J. Hist. Biol.* **9:**121–132.

Graunt, J., 1899, Natural and Political Observations Mentioned in a Following Index and Made upon the Bills of Mortality, in *The Economic Writings of Sir William Petty,* vol. 2, C. H. Hull, ed., Cambridge University Press, London.

Hardin, G., 1968, The Tragedy of the Commons, *Science* **162:**1243–1248. (This article appears as Paper 23 in *Eugenics Then and Now,* C. J. Bajema, ed., Benchmark Papers in Genetics, vol. 5, Dowden, Hutchinson & Ross, Stroudsburg, Pa., 1976, pp. 299–304.)

Hardin, G., 1974, Living on a Lifeboat, *BioScience,* **24:**561–568.

Hardin, G., 1977, *The Limits of Altruism: An Ecologist's View of Survival,* Indiana University Press, Bloomington, Ind., 154p.

Lawrence, W., 1819, *Lectures on Comparative Anatomy, Physiology, Zoology, and the Natural History of Man,* J. Calow, London, 579p. (An excerpt from the 1823 reissue is reprinted as Paper 8 in *Artificial Selection and the Development of Evolutionary Theory,* C. J. Bajema, ed., Benchmark Papers in Systematic and Evolutionary Biology, vol. 4, Hutchinson Ross, Stroudsburg, Pa., 1982, pp. 60–63.)

Malthus, T. R. (Anonymous), 1798, *An Essay on the Principle of Population as It Affects the Future Improvement of Society. With Remarks on the Speculations of Mr. Godwin, M. Condorcet and Other Writers,* 1st ed., Johnson, London, 396p. (reissue by St. Martin's Press, New York, 1966, 396p.)

Malthus, T. R., 1803, 1817, 1826, *An Essay on the Principle of Population; Or, A View of its Past and Present Effects on Human Happiness; with an Inquiry Into Our Prospects Respecting the Future Removal or Mitigation of the Evils Which It Occasions,* 2nd, 5th, and 6th eds., J. Murray, London, 610p.; 496p., 507p., 500p.; 535p., 528p.

Masters, R., 1968, *The Political Philosophy of Rousseau,* Princeton University Press, Princeton, N.J., 464p.

Masters, R., 1978, Jean-Jacques is Alive and Well: Rousseau and Contemporary Sociobiology, *Daedalus* **107:**93–105.

Ophuls, W., 1973, Leviathan or Oblivion? in *Toward a Steady State Economy,* H. Daley, ed., Freeman, San Francisco, pp. 215–230.

Ophuls, W., 1977, *Ecology and the Politics of Scarcity,* Freeman, San Francisco, 303p.

Passmore, J., 1970, *The Perfectibility of Man,* Scribner's, New York, 396p.

Passmore, J., 1974, *Man's Responsibility for Nature: Ecological Problems and Western Traditions,* Scribner's, New York, 213p.

Pearson, K., 1978, *The History of Statistics in the 17th and 18th Centuries,* Macmillan, New York, 744p.

Plattner, M., 1979, *Rousseau's State of Nature: An Interpretation of Discourse on Inequality,* Northern Illinois University Press, DeKalb, Ill., 137p.

Rousseau, J. J., 1755, *Discours sur l'Origine et les Fondements de l'Inegalite parmi les Hommes,* Geneva. (A 1761 translation has been reprinted in facsimile in *A Discourse Upon the Origin and Foundation of the Inequality Among Mankind,* Franklin, New York, 1971, 260p.)

Rousseau, J. J., 1762, *Du Contrat Social.* (A 1791 translation has been reprinted in facsimile in *The Social Contract,* C. Frankel, ed., Hafner, New York, 1962, 128p.)

Stocking, G. W., ed., 1973, *Researches into the Physical History of Man by J. C. Prichard,* University of Chicago Press, Chicago, 568p. (facsimile reprint of the 1813 edition published by Arch, London).

Walker, A., 1838, *Intermarriage: Or the Mode in Which, and the Causes Why, Beauty, Health and Intellect Result from Certain Unions and Deformity, Disease and Insanity from Others: Demonstrated by . . . and by an Account of Corresponding Effects in the Breeding of Animals,* J. Churchill, London.

Wallace, R. (Anonymous), 1761, *Various Prospects of Mankind, Nature and Providence,* 2nd ed., A. Millar, London, 406p. (reprinted in 1969 by Kelley, New York).

Weiss, K. M., and P. A. Ballonoff, eds., 1975, *Demographic Genetics,* Benchmark Papers in Genetics, vol. 3, Dowden, Hutchinson & Ross, Stroudsburg, Pa., 414p.

Wokler, R., 1978, Perfectible Apes in Decadent Cultures: Rousseau's Anthropology Revisited, *Daedalus* **107**:107–134.

1A

Reprinted from pages 60–63 of *Leviathan, or the Matter, Forme, & Power of a Common-wealth Ecclesiasticall and Civill*, Crooke, London, 1651

Of the NATURALL CONDITION *of Mankind, as concerning their* Felicity, *and* Misery.

T. Hobbes

Men by na-ture Equall.

NAture hath made men so equall, in the faculties of body, and mind; as that though there bee found one man sometimes manifestly stronger in body, or of quicker mind then another; yet when all is reckoned together, the difference between man, and man, is not so considerable, as that one man can thereupon claim to himselfe any benefit, to which another may not pretend, as well as he. For as to the strength of body, the weakest has strength enough to kill the strongest, either by secret machination, or by confederacy with others, that are in the same danger with himselfe.

And as to the faculties of the mind, (setting aside the arts grounded upon words, and especially that skill of proceeding upon generall, and infallible rules, called Science; which very few have, and but in few things; as being not a native faculty, born with us; nor attained, (as Prudence,) while we look after somewhat els,) I find yet a greater equality amongst men, than that of strength. For Prudence, is but Experience; which equall time, equally bestowes on all men, in those

those things they equally apply themselves unto. That which may perhaps make such equality incredible, is but a vain conceipt of ones owne wisdome, which almost all men think they have in a greater degree, than the Vulgar; that is, than all men but themselves, and a few others, whom by Fame, or for concurring with themselves, they approve. For such is the nature of men, that howsoever they may acknowledge many others to be more witty, or more eloquent, or more learned; Yet they will hardly believe there be many so wise as themselves: For they see their own wit at hand, and other mens at a distance. But this proveth rather that men are in that point equall, than unequall. For there is not ordinarily a greater signe of the equall distribution of any thing, than that every man is contented with his share.

From this equality of ability, ariseth equality of hope in the attaining of our Ends. And therefore if any two men desire the same thing, which neverthelesse they cannot both enjoy, they become enemies; and in the way to their End, (which is principally their owne conservation, and sometimes their delectation only,) endeavour to destroy, or subdue one an other. And from hence it comes to passe, that where an Invader hath no more to feare, than an other mans single power; if one plant, sow, build, or possesse a convenient Seat, others may probably be expected to come prepared with forces united, to dispossesse, and deprive him, not only of the fruit of his labour, but also of his life, or liberty. And the Invader again is in the like danger of another. *From Equality proceeds Diffidence.*

And from this diffidence of one another, there is no way for any man to secure himselfe, so reasonable, as Anticipation; that is, by force, or wiles, to master the persons of all men he can, so long, till he see no other power great enough to endanger him: And this is no more than his own conservation requireth, and is generally allowed. Also because there be some, that taking pleasure in contemplating their own power in the acts of conquest, which they pursue farther than their security requires; if others, that otherwise would be glad to be at ease within modest bounds, should not by invasion increase their power, they would not be able, long time, by standing only on their defence, to subsist. And by consequence, such augmentation of dominion over men, being necessary to a mans conservation, it ought to be allowed him. *From Diffidence Warre.*

Againe, men have no pleasure, (but on the contrary a great deale of griefe) in keeping company, where there is no power able to over-awe them all. For every man looketh that his companion should value him, at the same rate he sets upon himselfe: And upon all signes of contempt, or undervaluing, naturally endeavours, as far as he dares (which amongst them that have no common power to keep them in quiet, is far enough to make them destroy each other,) to extort a greater value from his contemners, by dommage; and from others, by the example.

So that in the nature of man, we find three principall causes of quarrell. First, Competition; Secondly, Diffidence; Thirdly, Glory.

The

The firſt, maketh men invade for Gain ; the ſecond, for Safety ; and the third, for Reputation. The firſt uſe Violence, to make themſelves Maſters of other mens perſons, wives, children, and cattell ; the ſecond, to defend them ; the third, for trifles, as a word, a ſmile, a different opinion, and any other ſigne of undervalue, either direct in their Perſons, or by reflexion in their Kindred, their Friends, their Nation, their Profeſſion, or their Name.

Out of Civil States, there is always Warre of every one against every one.

Hereby it is manifeſt, that during the time men live without a common Power to keep them all in awe, they are in that condition which is called Warre ; and ſuch a warre, as is of every man, againſt every man. For WARRE, conſiſteth not in Battell onely, or the act of fighting ; but in a tract of time, wherein the Will to contend by Battell is ſufficiently known : and therefore the notion of *Time*, is to be conſidered in the nature of Warre ; as it is in the nature of Weather. For as the nature of Foule weather, lyeth not in a ſhowre or two of rain ; but in an inclination thereto of many dayes together : So the nature of War, conſiſteth not in actuall fighting ; but in the known diſpoſition thereto, during all the time there is no aſſurance to the contrary. All other time is PEACE.

The Incommodities of such a War.

Whatſoever therefore is conſequent to a time of Warre, where every man is Enemy to every man ; the ſame is conſequent to the time, wherein men live without other ſecurity, than what their own ſtrength, and their own invention ſhall furniſh them withall. In ſuch condition, there is no place for Induſtry ; becauſe the fruit thereof is uncertain : and conſequently no Culture of the Earth ; no Navigation, nor uſe of the commodities that may be imported by Sea ; no commodious Building ; no Inſtruments of moving, and removing ſuch things as require much force ; no Knowledge of the face of the Earth ; no account of Time ; no Arts ; no Letters ; no Society ; and which is worſt of all, continuall feare, and danger of violent death ; And the life of man, ſolitary, poore, naſty, brutiſh, and ſhort.

It may ſeem ſtrange to ſome man, that has not well weighed theſe things ; that Nature ſhould thus diſſociate, and render men apt to invade, and deſtroy one another : and he may therefore, not truſting to this Inference, made from the Paſſions, deſire perhaps to have the ſame confirmed by Experience. Let him therefore conſider with himſelfe, when taking a journey, he armes himſelfe, and ſeeks to go well accompanied ; when going to ſleep, he locks his dores ; when even in his houſe he locks his cheſts ; and this when he knowes there bee Lawes, and publike Officers, armed, to revenge all injuries ſhall bee done him ; what opinion he has of his fellow ſubjects, when he rides armed ; of his fellow Citizens, when he locks his dores ; and of his children, and ſervants, when he locks his cheſts. Does he not there as much accuſe mankind by his actions, as I do by my words ? But neither of us accuſe mans nature in it. The Deſires, and other Paſſions of man, are in themſelves no Sin. No more are the Actions, that proceed from thoſe Paſſions, till they know a Law that forbids them : which till Lawes be made they cannot know : nor can any Law be made, till they have agreed upon the Perſon that ſhall make it.

It

It may peradventure be thought, there was never such a time, nor condition of warre as this; and I believe it was never generally so, over all the world: but there are many places, where they live so now. For the savage people in many places of *America*, except the government of small Families, the concord whereof dependeth on naturall lust, have no government at all; and live at this day in that brutish manner, as I said before. Howsoever, it may be perceived what manner of life there would be, where there were no common Power to feare; by the manner of life, which men that have formerly lived under a peacefull government, use to degenerate into, in a civill Warre.

But though there had never been any time, wherein particular men were in a condition of warre one against another; yet in all times, Kings, and Persons of Soveraigne authority, because of their Independency, are in continuall jealousies, and in the state and posture of Gladiators; having their weapons pointing, and their eyes fixed on one another; that is, their Forts, Garrisons, and Guns upon the Frontiers of their Kingdomes; and continuall Spyes upon their neighbours; which is a posture of War. But because they uphold thereby, the Industry of their Subjects; there does not follow from it, that misery, which accompanies the Liberty of particular men.

To this warre of every man against every man, this also is consequent; that nothing can be Unjust. The notions of Right and Wrong, Justice and Injustice have there no place. Where there is no common Power, there is no Law: where no Law, no Injustice. Force, and Fraud, are in warre the two Cardinall vertues. Justice, and Injustice are none of the Faculties neither of the Body, nor Mind. If they were, they might be in a man that were alone in the world, as well as his Senses, and Passions. They are Qualities, that relate to men in Society, not in Solitude. It is consequent also to the same condition, that there be no Propriety, no Dominion, no *Mine* and *Thine* distinct; but onely that to be every mans, that he can get; and for so long, as he can keep it. And thus much for the ill condition, which man by meer Nature is actually placed in; though with a possibility to come out of it, consisting partly in the Passions, partly in his Reason. *In such a Warre, nothing is Unjust.*

The Passions that encline men to Peace, are Feare of Death; Desire of such things as are necessary to commodious living; and a Hope by their Industry to obtain them. And Reason suggesteth convenient Articles of Peace, upon which men may be drawn to agreement. These Articles, are they, which otherwise are called the Lawes of Nature: whereof I shall speak more particularly, in the two following Chapters. *The Passions that incline men to Peace.*

1B

Reprinted from pages 64–66 of *Leviathan, or the Matter, Forme, & Power of a Common-wealth Ecclesiasticall and Civill*, Crooke, London, 1651

Of the first and second NATURALL LAWES, and of CONTRACTS.

T. Hobbes

Right of Nature what.

THe RIGHT OF NATURE, which Writers commonly call *Jus Naturale*, is the Liberty each man hath, to use his own power, as he will himselfe, for the preservation of his own Nature ; that is to say, of his own Life; and consequently, of doing any thing, which in his own Judgement, and Reason, hee shall conceive to be the aptest means thereunto.

Liberty what.

By LIBERTY, is understood, according to the proper signification of the word, the absence of externall Impediments : which Impediments, may oft take away part of a mans power to do what hee would ; but cannot hinder him from using the power left him, according as his judgement, and reason shall dictate to him.

A Law of Nature what.

A LAW OF NATURE, (*Lex Naturalis*,) is a Precept, or generall Rule, found out by Reason, by which a man is forbidden to do, that, which is destructive of his life, or taketh away the means of preserving the same ; and to omit, that, by which he thinketh it may be best preserved. For though they that speak of this subject, use to confound *Jus*, and *Lex*, *Right* and *Law* ; yet they ought to be distingui-

Difference of Right and Law.

shed ; because RIGHT, consisteth in liberty to do, or to forbeare ; Whereas LAw, determineth, and bindeth to one of them : so that Law, and Right, differ as much, as Obligation, and Liberty ; which in one and the same matter are inconsistent.

Naturally every man has Right to every thing.

And because the condition of Man, (as hath been declared in the precedent Chapter) is a condition of Warre of every one against every one ; in which case every one is governed by his own Reason ; and there is nothing he can make use of, that may not be a help unto him, in preserving his life against his enemyes ; It followeth, that in such a condition, every man has a Right to every thing ; even to one anothers body. And therefore, as long as this naturall Right of every man to every thing endureth, there can be no security to any man, (how strong or wise soever he be,) of living out the time, which Nature ordinarily alloweth men to live. And consequently it is a pre-

The Fundamentall Law of Nature.

cept, or generall rule of Reason, *That every man, ought to endeavour Peace, as farre as he has hope of obtaining it ; and when he cannot obtain it, that he may seek, and use, all helps, and advantages of Warre.* The first branch of which Rule, containeth the first, and Fundamentall Law of Nature; which is, *to seek Peace, and follow it.* The Second, the summe of the Right of Nature ; which is, *By all means we can, to defend our selves.*

The second Law of Nature.

From this Fundamentall Law of Nature, by which men are commanded to endeavour Peace, is derived this second Law ; *That a man be willing, when others are so too, as farre-forth, as for Peace, and*

defence

*defence of himselfe he shall think it necessary, to lay down this right to
all things; and be contented with so much liberty against other men,
as he would allow other men against himselfe.* For as long as every
man holdeth this Right, of doing any thing he liketh; so long are
all men in the condition of Warre. But if other men will not lay
down their Right, as well as he; then there is no Reason for any one,
to devest himselfe of his: For that were to expose himselfe to Prey,
(which no man is bound to) rather than to dispose himselfe to Peace.
This is that Law of the Gospell; *Whatsoever you require that others
should do to you, that do ye to them.* And that Law of all men, *Quod
tibi fieri non vis, alteri ne feceris.*

To *lay downe* a mans *Right* to any thing, is to *devest* himselfe of *What it is to lay down a Right.*
the *Liberty*, of hindring another of the benefit of his own Right to
the same. For he that renounceth, or passeth away his Right, giveth
not to any other man a Right which he had not before; because there
is nothing to which every man had not Right by Nature: but onely
standeth out of his way, that he may enjoy his own originall Right,
without hindrance from him; not without hindrance from another.
So that the effect which redoundeth to one man, by another mans
defect of Right, is but so much diminution of impediments to the use
of his own Right originall.

Right is layd aside, either by simply Renouncing it; or by Trans- *Renouncing a Right what it is.*
ferring it to another. By *Simply* RENOUNCING; when he cares *Transferring Right what.*
not to whom the benefit thereof redoundeth. By TRANSFERRING;
when he intendeth the benefit thereof to some certain person, or persons. And when a man hath in either manner abandoned, or granted
away his Right; then is he said to be OBLIGED, or BOUND, not to *Obligation.*
hinder those, to whom such Right is granted, or abandoned, from the
benefit of it: and that he *Ought*, and it is his DUTY, not to make *Duty.*
voyd that voluntary act of his own: and that such hindrance is IN-
IUSTICE, and INIURY, as being *Sine Jure*; the Right being be- *Injustice.*
fore renounced, or transferred. So that *Injury*, or *Injustice*, in the
controversies of the world, is somewhat like to that, which in the
disputations of Scholers is called *Absurdity*. For as it is there called
an Absurdity, to contradict what one maintained in the Beginning:
so in the world, it is called Injustice, and Injury, voluntarily to undo
that, which from the beginning he had voluntarily done. The way
by which a man either simply Renounceth, or Transferreth his
Right, is a Declaration, or Signification, by some voluntary and sufficient signe, or signes, that he doth so Renounce, or Transferre; or hath
so Renounced, or Transferred the same, to him that accepteth it. And
these Signes are either Words onely, or Actions onely; or (as it hap-
peneth most often) both Words, and Actions. And the same are
the BONDS, by which men are bound, and obliged: Bonds, that have
their strength, not from their own Nature, (for nothing is more easily broken then a mans word,) but from Feare of some evill consequence upon the rupture.

Whensoever a man Transferreth his Right, or Renounceth it; it *Not all Rights are alienable.*
is either in consideration of some Right reciprocally transferred to
himselfe;

17

himſelfe ; or for ſome other good he hopeth for thereby. For it is a voluntary act : and of the voluntary acts of every man, the object is ſome *Good to himſelfe*. And therefore there be ſome Rights, which no man can be underſtood by any words, or other ſignes, to have abandoned, or transferred. As firſt a man cannot lay down the right of reſiſting them, that aſſault him by force, to take away his life ; becauſe he cannot be underſtood to ayme thereby, at any Good to himſelfe. The ſame may be ſayd of Wounds, and Chayns, and Impriſonment ; both becauſe there is no benefit conſequent to ſuch patience ; as there is to the patience of ſuffering another to be wounded, or impriſoned : as alſo becauſe a man cannot tell, when he ſeeth men proceed againſt him by violence, whether they intend his death or not. And laſtly the motive, and end for which this renouncing, and transferring of Right is introduced, is nothing elſe but the ſecurity of a mans perſon, in his life, and in the means of ſo preſerving life, as not to be weary of it. And therefore if a man by words, or other ſignes, ſeem to deſpoyle himſelfe of the End, for which thoſe ſignes were intended ; he is not to be underſtood as if he meant it, or that it was his will ; but that he was ignorant of how ſuch words and actions were to be interpreted.

Contract
what.
    The mutuall transferring of Right, is that which men call Co n-
t r a c t.

    There is difference, between transferring of Right to the Thing ; and transferring, or tradition, that is, delivery of the Thing it ſelfe. For the Thing may be delivered together with the Tranſlation of the Right ; as in buying and ſelling with ready mony ; or exchange of goods, or lands : and it may be delivered ſome time after.

    Again, one of the Contractors, may deliver the Thing contracted for on his part, and leave the other to perform his part at ſome determinate time after, and in the mean time be truſted ; and then the Con-
Covenant
what.
tract on his part, is called P a c t, or C o v e n a n t : Or both parts may contract now, to performe hereafter : in which caſes, he that is to performe in time to come, being truſted, his performance is called *Keeping of Promiſe*, or Faith ; and the fayling of performance (if it be voluntary) *Violation of Faith*.

    When the transferring of Right, is not mutuall ; but one of the parties transferreth, in hope to gain thereby friendſhip, or ſervice from another, or from his friends ; or in hope to gain the reputation of Charity, or Magnanimity ; or to deliver his mind from the pain of compaſſion ; or in hope of reward in heaven ; This is not Contract,
Free-gift.
but G i f t, Free-gift, Grace : which words ſignifie one and the ſame thing.

[*Editor's Note:* Material has been omitted at this point.]

2A

Reprinted from pages 193–195 of *The Works of William Harvey, M.D.,* Sydenham Society, London, 1847, 624p.

OF THE UTERUS OF THE FOWL

W. Harvey

[*Editor's Note:* In the original, material precedes this excerpt.]

A male pheasant kept in an aviary was so inflamed with lust, that unless he had the company of several hen-birds, six at the least, he literally maltreated them, though his repeated addresses rather interfered with their breeding than promoted it. I have seen a single hen-pheasant shut up with a cock-bird (which she could in no way escape) so worn out, and her back so entirely stript of feathers through his reiterated assaults, that at length she died exhausted. In the body of this bird, however, I did not discover even the rudiments of eggs.

I have also observed a male duck, having none of his own kind with him, but associating with hens, inflamed with such desire that he would follow a pullet even for several hours, would seize her with his bill, and mounting at length upon the creature, worn out with fatigue, would compel her to submit to his pleasure.

The common cock, victorious in a battle, not only satisfies his desires upon the sultanas of the vanquished, but upon the body of his rival himself.

The females of some animals are likewise so libidinous that they excite their males by pecking or biting them gently about the head; they seem as if they whispered into their ears the sweets of love; and then they mount upon their backs and invite them by other arts to fruition: among the number may be mentioned pigeons and sparrows.

It did not therefore appear likely that a few treads, in the beginning of the year, should suffice to render fertile the whole of the eggs that are to be laid in its course.

Upon one occasion, however, in the spring season, by way of helping out Fabricius, and that I might have some certain data as to the time during which the fecundating influence of intercourse would continue, and the necessity of renewed communication, I had a couple of hens separated from the cock for four days, each of which laid three eggs, all of which were prolific. Another hen was secluded, and the egg she laid on the tenth day afterwards was fruitful. The egg which another laid on the twentieth day of her seclusion also produced a chick. It would therefore seem that intercourse, once or twice repeated, suffices to impregnate the whole bunch of yelks, the whole of the eggs that will be laid during a certain season.

I shall here relate another observation which I made at this time. When I returned two of the hens, which I had secluded for a time, to the cock, one of which was big with egg, the other having but just laid, the cock immediately ran to the latter and trod her greedily three or four times; the former he went round and round, tripping himself with his wing and seeming to salute her, and wish her joy of her return; but he soon returned to the other and trod her again and again, even compelling her to submit; the one big with egg, however, he always speedily forsook, and never solicited her to his pleasure. I wondered with myself by what signs he knew that intercourse would advantage one of these hens and prove unavailing to the other. But indeed it is not easy at any time to understand how male animals, even from a distance, know which females are in season and desirous of their company; whether it be by sight, or hearing, or smell, it is difficult to say. Some on merely hearing the voice of the female, or smelling at the place where she has made water, or even the ground over which she has passed, are straightway seized with desire and set off in pursuit to gratify it. But I shall have more to say on this subject in my treatise on the Loves, Lusts, and Sexual Acts of Animals. I return to the matter we have in hand.

2B

Reprinted from pages 311–312 of *The Works of William Harvey, M.D.,* Sydenham Society, London, 1847, 624p.

OF THE COCK AND THE PARTICULARS MOST REMARKABLE IN HIS CONSTITUTION

W. Harvey

[*Editor's Note:* In the original, material precedes this excerpt.]

Among male animals there is none that is more active or more haughty and erect, or that has stronger powers of digestion than the cock, which turns the larger portion of his food into semen; hence it is that he requires so many wives,— ten or even a dozen. For there are some animals, single males of which suffice for several females, as we see among deer, cattle, &c.; and there are others, of which the females are so prurient that they are scarcely satisfied with several males, such as the bitch and the wolf; whence prostitutes were called *lupæ* or wolves, as making their persons common; and stews were entitled *lupanaria*. Whilst some animals, of a more chaste disposition, live, as it were, in the conjugal estate, so that the male is married to a single female only, and both take part in providing for the wants of the family; for since nature requires that the male supply the deficiencies of the female in the work of generation, and as she alone in many cases does not suffice to cherish and feed and protect the young, the male is added to the wife that he may take part in the burthen of bringing up the offspring. Partridges lead a wedded life, because the females alone cannot incubate such a number of eggs as they lay, (so that they are said, by some, to make two nests,) nor to bring up such a family as by and by appears without assistance. The male pigeon also assists in building the nest, takes his turn in incubating the eggs, and is active in feeding the young. In the same way many other instances of conjugal life among the lower animals might be quoted, and indeed we shall have occasion to refer to several in what yet remains to be said.

Those males, among animals, which serve several females, such as the cock, have an abundant secretion of seminal fluid, and are provided with long and ample vasa deferentia. And at whatever time or season the clustered rudimentary papulæ in the ovary come to maturity and require fecundation, that they may go on to be turned into perfect eggs, the males will then be found to have an abundance of seminal fluid, and the testicles to enlarge and become conspicuous in the very situation to which they transfer their fecundating influence, viz. the præcordia.   This is remarkable in fishes, birds, and the whole race of oviparous animals; the males of which teem with fecundating seminal fluid at the same precise seasons as the females become full of eggs.

Whatever parts of the hen, therefore, are destined by nature for purposes of generation, viz. the ovary, the infundibulum, the processus uteri, the uterus itself, and the pudenda; as also the situation of these parts, their structure, dimensions, temperature, and all that follows this; all these, I say, are either subordinate to the production and growth of the egg, or to intercourse and the reception of fecundity from the male; or, for the sake of parturition, to which they conduce either as principal and convenient means, or as means necessary, and without which what is done could not be accomplished; for nothing in nature's works is fashioned either carelessly or in vain.   In the same way all the parts in the cock are fashioned subordinate to the preparation or concoction of the spermatic fluid, and its transference to the hen.

Now those males that are so vigorously constituted as to serve several females are larger and handsomer, and in the matter of spirit and arms excel their females in a far greater degree than the males of those that live attached to a single female.   Neither the male partridge, nor the crow, nor the pigeon, is distinguished from the female bird in the same decided way as the cock from his hens, the stag from his does, &c.

The cock, therefore, as he is gayer in his plumage, better armed, more courageous and pugnacious, so is he replete with semen, and so apt for repeated intercourse, that unless he have a number of wives he distresses them by his frequent assaults; he not only invites but compels them to his pleasure, and leaping upon them at inconvenient and improper seasons, (even when they are engaged in the business of incubation) and wearing off the feathers from their backs, he truly does them an injury. I have occasionally seen hens so torn and worn by the ferocious addresses of the cock, that with their backs stript of feathers and laid bare in places, even to the bone, they languished miserably for a time and then died.   The same thing also occurs among pheasants, turkeys, and other species.

2c

Reprinted from pages 425–426 of *The Works of William Harvey, M.D.*, Sydenham Society, London, 1847, 624p.

OF THE ORDER OF THE PARTS IN GENERATION AS IT APPEARS FROM OBSERVATION

W. Harvey

[*Editor's Note:* In the original, material precedes this excerpt.]

Man comes into the world naked and unarmed, as if nature had destined him for a social creature, and ordained him to live under equitable laws and in peace; as if she had desired that he should be guided by reason rather than be driven by force; therefore did she endow him with understanding, and furnish him with hands, that he might himself contrive what was necessary to his clothing and protection. To those animals to which nature has given vast strength, she has also presented weapons in harmony with their powers; to those that are not thus vigorous, she has given ingenuity, cunning, and singular dexterity in avoiding injury.

Ornaments of all kinds, such as tufts, crests, combs, wattles, brilliant plumage, and the like, of which some vain creatures seem not a little proud, to say nothing of such offensive weapons as teeth, horns, spurs, and other implements employed in combat, are more frequently and remarkably conferred upon the male than the female. And it is not uninteresting to remark, that many of these ornaments or weapons are most conspicuous in the male at that epoch when the females come into season, and burn with desire of engendering. And whilst in the young they are still absent, in the aged they also fail as being no longer wanted.

Our common cock, whose pugnacious qualities are well known, so soon as he comes to his strength and is possessed of the faculty of engendering, is distinguished by his spurs, and ornamented with his comb and beautiful feathers, by which he charms his mates to the rites of Venus, and is furnished for the combat with other males, the subject of dispute being no empty or vainglorious matter, but the perpetuation of the stock in this line or in that; as if nature had intended that he who could best defend himself and his, should be preferred to others for the continuance of the kind. And indeed all animals which are better furnished with weapons of offence, and more warlike than others, fall out and fight, either in defence of their young, of their nests or dens, or of their prey; but more than all for the possession of their females. Once vanquished, they yield up possession of these, lay aside their strut and haughty demeanour, and, crest-fallen and submissive, they seem to consume with grief; the victor, on the contrary, who has gained possession of the females by his prowess, exults and boastfully proclaims the glory of his conquest.

Nor is this ornamenting anything adventitious and for a season only; it is a lasting and special gift of nature, who has not been studious to deck out animals, and especially birds only, but has also thrown an infinite variety of beautiful dyes over the lowly and insensate herbs and flowers.

3

OF THE DIFFERENCE BETWEEN THE NUMBERS
OF MALES AND FEMALES

J. Graunt

[*Editor's Note:* In the original, material precedes and follows this excerpt.]

THE next Observation is, That there be more *Males* than *Females*[1].

1. There have been Buried from the year 1628, to the year 1662, *exclusive,* 209436 *Males,* and but 190474 *Females:* but it will be objected, That in *London* it may be indeed so, though otherwise elsewhere; because *London* is the great Stage and Shop of business, wherein the *Masculine Sex* bears the greatest part. But we Answer, That there have been also *Christened* within the same time 139782 *Males,* and (65) but 130866 *Females,* and that ‖ the Country-Accounts are consonant enough to those of *London* upon this matter[2].

2. What the Causes hereof are, we shall not trouble our selves to conjecture, as in other Cases: only we shall desire that Travellers would enquire, whether it be the same in other Countries.

3. We should have given an Account, how in every Age these proportions change here, but that we have Bills of distinction but for 32 years, so that we shall pass from hence to some Inferences from this Conclusion; as first,

I. That *Christian Religion,* prohibiting *Polygamy,* is more agreeable to the *Law of Nature,* that is, the *Law of God,* than *Mahumetism,* and others, that allow it: for one Man his having many Women, or Wives, by Law, signifies nothing, unless there were many Women to one Man in Nature also.

---

[1] The Table of Males and Females is at p. 411.   [2] See p. 389.

II. The obvious Objection hereunto is, That one *Horse*, *Bull*, or *Ram*, having each of them many *Females*, do promote increase. To which I Answer, That although perhaps there be naturally, even of these *species*, more *Males* than *Females*, yet *artificially*, that is, by making *Geldings*, *Oxen*, and *Weathers*, there are fewer. From whence it will follow, That when by experience it is found how ma-‖ny *Ews* (suppose twenty) (66) one *Ram* will serve, we may know what proportion of *male-Lambs* to castrate or geld, *viz.* nineteen, or thereabouts : for if you emasculate fewer, *viz.* but ten, you shall, by promiscuous copulation of each of those ten with two *Females*, hinder the increase, so far as the admittance of two *Males* will do it : but, if you castrate none at all, it is highly probable, that, every of the twenty *Males* copulating with every of the twenty *Females*, there will be little or no conception in any of them all.

III. And this I take to be the truest Reason, why *Foxes*, *Wolves*, and other *Vermin Animals*, that are not gelt, increase not faster than *Sheep*, when as so many thousands of these are daily Butchered, and very few of the other die otherwise than of themselves.

4. We have hitherto said, There are more *Males* than *Females*; we say next, That the one exceed the other by about a thirteenth part. So that although more Men die violent deaths than Women, that is, more are *slain* in *Wars*, *killed* by *Mischance*, *drowned* at *Sea*, and die by the *Hand of Justice*; moreover, more Men go to *Colonies*, and travel into Forein parts, than Women ; and lastly, more remain unmarried than of Women, as *Fellows* of *Colleges*, and *Apprentices* above eighteen, ‖ *&c.* yet the said thirteenth (67) part difference bringeth the business but to such a pass, that every Woman may have an Husband, without the allowance of *Polygamy*.

5. Moreover, although a Man be *Prolifick* fourty years, and a Woman but five and twenty, which makes the *Males* to be as 560 to 325 *Females*, yet the causes above-named, and the later marriage of the Men, reduce all to an equality.

6. It appearing, that there were fourteen Men to thirteen

Women, and that they die in the same proportion also; yet I have heard *Physicians* say, that they have two Women Patients to one Man, which Assertion seems very likely; for that Women have either the *Green-sickness*, or other like Distempers, are sick of *Breedings, Abortions, Child-bearing, Sore-breasts, Whites, Obstructions, Fits of the Mother*, and the like.

7. Now from this it should follow, that more Women should die than Men, if the number of *Burials* answered in proportion to that of Sicknesses: but this must be salved, either by the alleging, that the *Physicians* cure those Sicknesses, so as few more die than if none were sick; or else that Men, being more intemperate than Women, die as much (68) by reason of their Vices, as Women do by the Infir-||mity of their *Sex*; and consequently, more *Males* being born than *Females*, more also die.

8. In the year 1642 many *Males* went out of *London* into the Wars then beginning, insomuch as I expected in the succeeding year 1643 to have found the *Burials* of *Females* to have exceeded those of *Males*, but no alteration appeared; forasmuch, as I suppose, Trading continuing the same in *London*, all those, who lost their *Apprentices*, had others out of the Country; and if any left their Trades and Shops, that others forthwith succeeded them: for, if employment for hands remain the same, no doubt but the number of them could not long continue in disproportion.

9. Another pregnant Argument to the same purpose (which hath already been touched on) is, That although in the very year of the *Plague* the *Christenings* decreased, by the dying and flying of *Teeming-Women*, yet the very next year after they increased somewhat, but the second after to as full a number as in the second year before the said *Plague*: for I say again, if there be encouragement for an hundred in *London*, that is, a Way how an hundred may live better than in the Country, and if there be void Housing there to receive|| (69) them, the evacuating of a fourth or third part of that number must soon be supplied out of the Country; so as the great *Plague* doth not lessen the Inhabitants of the City, but of

the Country, who in a short time remove themselves from thence hither, so long, until the City, for want of receipt and encouragement, regurgitates and sends them back.

10. From the difference between *Males* and *Females*, we see the reason of making *Eunuchs* in those places where *Polygamy* is allowed, the later being useless as to multiplication, without the former, as was said before in case of *Sheep* and other *Animals* usually gelt in these Countries.

11. By consequence, this practice of *Castration* serves as well to promote increase, as to meliorate the Flesh of those Beasts that suffer it. For that Operation is equally practised upon *Horses*, which are not used for food, as upon those that are.

12. In *Popish* Countries, where *Polygamy* is forbidden, if a greater number of *Males* oblige themselves to *Cælibate*, than the natural over-plus, or difference between them and *Females* amounts unto; then multiplication is hindred: for if there be eight Men to ten Women, all of which eight Men are married to eight of the ten Women, then the other two ‖ bear no Children, as either admitting no Man at all, or else (70) admitting Men as Whores (that is, more than one;) which commonly procreates no more than if none at all had been used: or else such unlawful Copulations beget Conceptions, but to frustrate them by procured Abortions, or secret Murthers; all which returns to the same reckoning. Now, if the same proportion of Women oblige themselves to a single life likewise, then such obligation makes no change in this matter of increase.

13. From what hath been said appears the reason, why the Law is and ought to be so strict against Fornications and Adulteries: for, if there were universal liberty, the Increase of Mankind would be but like that of *Foxes* at best.

14. Now forasmuch as Princes are not only Powerful, but Rich, according to the number of their People (Hands being the Father, as Lands are the Mother and Womb of Wealth)[1] it is no wonder why States, by encouraging Marriage,

---

[1] This idea, which occurs in slightly different phraseology in Petty's *Treatise of Taxes* (p. 68), has been pronounced a "leading thought in his writings."

and hindering Licentiousness, advance their own Interest, as well as preserve the Laws of God from contempt and violation.

15. It is a Blessing to Mankind, that by this over-plus (71) of *Males* there is this natural ‖ Bar to *Polygamy*: for in such a state Women could not live in that parity and equality of expense with their Husbands, as now, and here they do.

16. The reason whereof is, not, that the Husband cannot maintain as splendidly three, as one; for he might, having three Wives, live himself upon a quarter of his Income, that is, in a parity with all three, as well as, having but one, live in the same parity at half with her alone: but rather, because that to keep them all quiet with each other, and himself, he must keep them all in greater aw, and less splendour; which power he having, he will probably use it to keep them all as low as he pleases, and at no more cost than makes for his own pleasure; the poorest Subjects, (such as this plurality of Wives must be) being most easily governed. ‖

Ingram, *Hist. of Political Economy*, 51; the suggestion is followed by Bevan, *Sir W. Petty, a Study*, 53. The figure in which the idea is expressed apparently reflects the current notion, at least as old as Aristotle, that the female is passive in generation. Legouvé, *Moral history of Woman*, tr. Palmer, 216. Even the form of expressing the analogy is, probably, older than either Graunt or Petty, for both place the words in brackets—a seventeenth century equivalent for marks of quotation—and Schulz, in his translation of Graunt, writes, "weil, nach dem Sprichwort, die hander der welt vater, und das land derselbten mutter ist."

4

Reprinted from *R. Soc. (London) Philos. Trans.* **27**:186–190 (1710)

II. *An Argument for Divine Providence, taken from the constant Regularity observ'd in the Births of both Sexes.* By *Dr.* John Arbuthnott, *Physitian in Ordinary to Her Majesty, and Fellow of the College of Physitians and the Royal Society.*

A Mong innumerable Footsteps of Divine Providence to be found in the Works of Nature, there is a very remarkable one to be observed in the exact Ballance that is maintained, between the Numbers of Men and Women; for by this means it is provided, that the Species may never fail, nor perish, since every Male may have its Female, and of a proportionable Age. This Equality of Males and Females is not the Effect of Chance but Divine Providence, working for a good End, which I thus demonstrate:

Let there be a Die of Two sides, M and F, (which denote Cross and Pile), now to find all the Chances of any determinate Number of such Dice, let the Binome M+F be raised to the Power, whose Exponent is the Number of Dice given; the Coefficients of the Terms will shew all the Chances sought. For Example, in Two Dice of Two sides M+F the Chances are $M^2+2MF+F^2$, that is, One Chance for M double, One for F double, and Two for M single and F single; in Four such Dice there are Chances $M^4+4M^3F+6M^2F^2+4MF^3+F^4$, that is, One Chance for M quadruple, One for F quadruple, Four for triple M and single F, Four for single M and triple F, and Six for M double and F double; and universally, if the Number of Dice be n, all their Chances will be expressed in this Series

$$M^n+$$

$$M^0 + \tfrac{n}{1} \times M^{n-1}F + \tfrac{n}{1} \times \tfrac{n-1}{2} \times M^{n-2}F^2 + \tfrac{n}{1} \times \tfrac{n-1}{2} \times \tfrac{n-1}{3} \times M^{n-3}F^3 +, \&c.$$

It appears plainly, that when the Number of Dice is even there are as many M's as F's in the middle Term of this Series, and in all the other Terms there are most M's or most F's.

If therefore a Man undertake with an even Number of Dice to throw as many M's as F's, he has all the Terms but the middle Term against him ; and his Lot is to the Sum of all the Chances, as the coefficient of the middle Term is to the power of 2 raised to an exponent equal to the Number of Dice: so in Two Dice his Lot is $\frac{2}{4}$ or $\frac{1}{2}$, in Three Dice $\frac{6}{8}$ or $\frac{3}{4}$, in Six Dice $\frac{20}{64}$ or $\frac{5}{16}$, in Eight $\frac{70}{128}$ or $\frac{35}{64}$, $\&c.$

To find this middle Term in any given Power or Number of Dice, continue the Series $\frac{n}{1} \times \frac{n-1}{2} \times \frac{n-1}{3}$, $\&c.$ till the number of terms are equal to $\frac{1}{2}n$. For Example, the coefficient of the middle Term of the tenth Power is $\frac{10}{1} \times \frac{9}{2} \times \frac{8}{3} \times \frac{7}{4} \times \frac{6}{5} = 252$, the tenth Power of 2 is 1024, if therefore A undertakes to throw with Ten Dice in one throw an equal Number of M's and F's, he has 252 Chances out of 1024 for him, that is his Lot is $\frac{252}{1024}$ or $\frac{63}{256}$, which is less than $\frac{1}{4}$.

It will be easy by the help of Logarithms, to extend this Calculation to a very great Number, but that is not my present Design. It is visible from what has been said, that with a very great Number of Dice, A's Lot would become very small; and consequently (supposing M to denote Male and F Female) that in the vast Number of Mortals, there would be but a small part of all the possible Chances, for its happening at any assignable time, that an equal Number of Males and Females should be born.

It is indeed to be confessed that this Equality of Males and Females is not Mathematical but Physical, which alters much the foregoing Calculation ; for in this Case the

the middle Term will not exactly give A's Chances, but his Chances will take in some of the Terms next the middle one, and will lean. to one side or the other. But it is very improbable (if mere Chance govern'd) that they would never reach as far as the Extremities: But this Event is wisely prevented by the wise Oeconomy of Nature; and to judge of the wisdom of the Contrivance, we must observe that the external Accidents to which Males are subject (who must seek their Food with danger) do make a great havock of them, and that this loss exceeds far that of the other Sex, occasioned by Diseases incident to it, as Experience convinces us. To repair that Loss, provident Nature, by the Disposal of its wise Creator, brings forth more Males than Females; and that in almost a constant proportion. This appears from the annexed Tables, which contain Observations for 82 Years of the Births in *London*. Now, to reduce the Whole to a Calculation, I propose this

Problem. A lays against B, that every Year there shall be born more Males than Females: To find A's Lot, or the Value of his Expectation.

It is evident from what has been said, that A's Lot for each Year is less than $\frac{1}{2}$; (but that the Argument may be stronger) let his Lot be equal to $\frac{1}{2}$ for one Year. If he undertakes to do the same thing 82 times running, his Lot will be $\overline{\frac{1}{2}}|^{82}$, which will be found easily by the Table of Logarithms to be $\frac{1}{4\,8360\ 0000\ 0000\ 0000\ 0000\ 0000}$. But if A wager with B, not only that the Number of Males shall exceed that of Females, every Year, but that this Excess shall happen in a constant Proportion, and the Difference lye within fix'd limits; and this not only for 82 Years, but for Ages of Ages, and not only at *London*, but all over the World; (which 'tis highly probable is Fact, and designed that every Male may have a Female of the same Country and suitable Age) then A's Chance will be near an infinitely small Quantity, at least less

leſs than any aſſignable Fraction. From whence it follows, that it is Art, not Chance, that governs.

There ſeems no more probable Cauſe to be aſſigned in Phyſicks for this Equality of the Births, than that in our firſt Parents Seed there were at firſt ſormed an equal Number of both Sexes.

Scholium. From hence it follows, that Polygamy is contrary to the Law of Nature and Juſtice, and to the Propagation of Human Race; for where Males and and Females are in equal number, if one Man takes Twenty Wives, Nineteen Men muſt live in Celibacy, which is repugnant to the Deſign of Nature; nor is it probable that Twenty Women will be ſo well impregnated by one Man as by Twenty.

Chriſtened.			Chriſtened.		
Anno.	*Males*	*Females.*	*Anno.*	*Males.*	*Females.*
1629	5218	4683	1648	3363	3181
30	4858	4457	49	3079	2746
31	4422	4102	50	2890	2722
32	4994	4590	51	3231	2840
33	5158	4839	52	3220	2908
34	5035	4820	53	3196	2959
35	5106	4928	54	3441	3179
36	4917	4605	55	3655	3349
37	4703	4457	56	3668	3382
38	5359	4952	57	3396	3289
39	5366	4784	58	3157	3013
40	5518	5332	59	3209	2781
41	5470	5200	60	3724	3247
42	5460	4910	61	4748	4107
43	4793	4617	62	5216	4803
44	4107	3997	63	5411	4881
45	4047	3919	64	6041	5681
46	3768	3395	65	5114	4858
47	3796	3536	66	4678	4319

B b   Chriſtened.

34

	Christened.				Christened.	
Anno.	Males.	Females.	Anno.	Males.	Females.
1667	5616	5322	1689	7604	7167
68	6073	5560	90	7909	7302
69	6506	5829	91	7662	7392
70	6278	5719	92	7602	7316
71	6449	6061	93	7676	7483
72	6443	6120	94	6985	6647
73	6073	5822	95	7263	6713
74	6113	5738	96	7632	7229
75	6058	5717	97	8062	7767
76	6552	5847	98	8426	7626
77	6423	6203	99	7911	7452
78	6568	6033	1700	7578	7061
79	6247	6041	1701	8102	7514
80	6548	6299	1702	8031	7656
81	6822	6533	1703	7765	7683
82	6909	6744	1704	6113	5738
83	7577	7158	1705	8366	7779
84	7575	7127	1706	7952	7417
85	7484	7246	1707	8379	7687
86	7575	7119	1708	8239	7623
87	7737	7214	1709	7840	7380
88	7487	7101	1710	7640	7288

5

Reprinted from pages 68–94 of *A Discourse Upon the Origin and Foundation of the Inequality Among Mankind*, 1761, London

A DISCOURSE UPON THE ORIGIN AND FOUNDATION OF THE INEQUALITY AMONG MANKIND

J. Rousseau

[*Editor's Note:* In the original, material precedes and follows this excerpt.]

But above all things let us beware concluding with *Hobbes*, that Man, as having no Idea of Goodnefs, muft be naturally bad; that he is vicious becaufe he does not know what Virtue is; that he always refufes to do any Service to thofe of his own Species, becaufe he believes that none is due to them; that, in virtue of that Right which he juftly claims to every thing he wants, he foolifhly looks upon himfelf as Proprietor of the whole Univerfe. *Hobbes* very plainly faw the Flaws in all the modern Definitions of Natural Right: but the Confequences, which he draws from his own Definition, fhow that it is, in the Senfe he underftands it, equally exceptionable. This Author, to argue from his own Principles, fhould fay that the State of Nature, being that

where

where the Care of our own Preſervation interferes leaſt with the Preſervation of others, was of courſe the moſt favourable to Peace, and moſt ſuitable to Mankind; whereas he advances the very reverſe in conſequence of his having injudiciouſly admitted, as Objeᴄts of that Care which Savage Man ſhould take of his Preſervation, the Satisfaᴄtion of numberleſs Paſ-ſions which are the work of Society, and have rendered Laws neceſſary. A bad Man, ſays he, is a robuſt Child. But this is not proving that Savage Man is a robuſt Child; and tho' we were to grant that he was, what could this Philoſopher infer from ſuch a Conceſſion? That if this Man, when robuſt, depended on others as much as when feeble, there is no Exceſs that he would not be guilty of. He would make nothing of ſtriking his Mother when ſhe delayed ever ſo little to give him the Breaſt; he would claw, and bite, and ſtrangle without remorſe

the

the firſt of his younger Brothers, that ever
ſo accidentally joſtled or otherwiſe diſturbed
him. But theſe are two contradictory Sup-
poſitions in the State of Nature, to be
robuſt and dependent. Man is weak when
dependent, and his own Maſter before he
grows robuſt. *Hobbes* did not conſider
that the ſame Cauſe, which hinders Savages
from making uſe of their Reaſon, as our
Juriſconſults pretend, hinders them at
the ſame time from making an ill uſe of
their Faculties, as he himſelf pretends ;
ſo that we may ſay that Savages are not
bad, preciſely becauſe they don't know
what it is to be good ; for it is neither
the Development of the Underſtanding,
nor the Curb of the Law, but the Calm-
neſs of their Paſſions and their Ignorance
of Vice that hinders them from doing ill :
tanto plus in illis proficit Vitiorum ignoran-
tia, quam in his cognitio Virtutis. There
is beſides another Principle that has eſcap-
ed *Hobbes,* and which, having been given
to

to Man to moderate, on certain Occafions, the blind and impetuous Sallies of
Self-love, or the Defire of Self-prefer-
vation previous to the Appearance of that
Paffion, (15) allays the Ardour, with which
he naturally purfues his private Welfare,
by an innate Abhorrence to fee Beings
fuffer that refemble him. I fhall not
furely be contradicted, in granting to Man
the only natural Virtue, which the moft
paffionate Detractor of human Virtues
could not deny him, I mean that of Pity,
a Difpofition fuitable to Creatures weak
as we are, and liable to fc many Evils;
a Virtue fo much the more univerfal, and
withal ufeful to Man, as it takes place in
him of all manner of Reflection; and fo
natural, that the Beafts themfelves fometimes give evident Signs of it. Not to
fpeak of the Tendernefs of Mothers for
their Young; and of the Dangers they
face to fcreen them from Danger; with
what Reluctance are Horfes known to
trample

39

trample upon living Bodies; one Animal never paſſes unmoved by the dead Carcaſs of another Animal of the ſame Species: there are even ſome who beſtow a kind of Sepulture upon their dead Fellows; and the mournful Lowings of Cattle, on their entering the Slaughter Houſe, publiſh the Impreſſion made upon them by the horrible Spectacle they are there ſtruck with. It is with Pleaſure we ſee the Author of the Fable of the Bees, forced to acknowledge Man a compaſſionate and ſenſible Being; and lay aſide, in the Example he offers to confirm it, his cold and ſubtile Stile, to place before us the pathetic Picture of a Man, who, with his Hands tied up, is obliged to behold a Beaſt of Prey tear a Child from the Arms of his Mother, and then with his Teeth grind the tender Limbs, and with his Claws rend the throbbing Entrails of the innocent Victim. What horrible Emotions muſt not ſuch a Spectator

tator experience at the fight of an Event which does not perfonally concern him? What anguifh muft he not fuffer at his not being able to affift the fainting Mother or the expiring Infant?

Such is the pure Motion of Nature, anterior to all manner of Reflection; fuch is the Force of natural Pity, which the moft diffolute Manners have as yet found it fo difficult to extinguifh, fince we every Day fee, in our theatrical Reprefentations, thofe Men fympathize with the unfortunate and weep at their Sufferings, who, if in the Tyrant's Place, would aggravate the Torments of their Enemies. *Mandeville* was very fenfible that Men, in fpite of all their Morality, would never have been better than Monfters, if Nature had not given them Pity to affift Reafon: but he did not perceive that from this Quality alone
flow

flow all the focial Virtues, which he would difpute Mankind the Poffeffion of. In fact, what is Generofity, what Clemency, what Humanity, but Pity applied to the Weak, to the Guilty, or to the Human Species in general ? Even Benevolence and Friendfhip, if we judge right, will appear the Effects of a conftant Pity, fixed upon a particular Object : for to wifh that a Perfon may not fuffer, what is it but to wifh that he may be happy ? Tho' it were true that Commiferation is no more than a Sentiment, which puts us in the place of him who fuffers, a Sentiment obfcure but active in the Savage, developed but dormant in civilized Man, how could this Notion affect the Truth of what I advance, but to make it more evident. In Fact, Commiferation muft be fo much the more energetic, the more intimately the Animal, that beholds any kind of Diftrefs, identifies himfelf with

the

the Animal that labours under it. Now
it is evident that this Identification muſt
have been infinitely more perfect in the
State of Nature, than in the State of
Reaſon. It is Reaſon that engenders
Self-love, and Reflection that ſtrengthens
it; it is Reaſon that makes Man ſhrink into
himſelf; it is Reaſon that makes him keep
aloof from every thing that can trouble
or afflict him : it is Philoſophy that de-
ſtroys his Connections with other Men;
it is in conſequence of her Dictates that he
mutters to himſelf at the ſight of another
in Diſtreſs, You may periſh for aught I
care, nothing can hurt me. Nothing
leſs than thoſe Evils, which threaten
the whole Species, can diſturb the calm
Sleep of the Philoſopher, and force him
from his Bed. One Man may with Im-
punity murder another under his Win-
dows; he has nothing to do but clap
his Hands to his Ears, argue a little with
himſelf to hinder Nature, that ſtartles
within

within him, from identifying him with
the unhappy Sufferer. Savage Man wants
this admirable Talent; and for want of
Wifdom and Reafon, is always ready
foolifhly to obey the firft Whifpers of
Humanity. In Riots and Street-Brawls
the Populace flock together, the prudent
Man fneaks off. They are the Dregs of
the People, the poor Bafket and Barrow-
women, that part the Combatants, and
hinder gentle Folks from cutting one an-
ther's Throats.

It is therefore certain that Pity is a natural Sentiment, which, by moderating in
every Individual the Activity of Self-love,
contributes to the mutual Prefervation of
the whole Species. It is this Pity which
hurries us without Reflection to the Af-
fiftance of thofe we fee in Diftrefs; it is
this Pity which, in a State of Nature,
ftands for Laws, for Manners, for Virtue,
with this Advantage, that no one is
tempted

tempted to difobey her fweet and gentle
Voice: it is this Pity which will always
hinder a robuft Savage from plundering
a feeble Child, or infirm old Man, of
the Subfiftence they have acquired with
Pain and Difficulty, if he has but the leaft
Profpect of providing for himfelf by any
other Means: it is this Pity which, inftead of that fublime Maxim of argumen-
tative Juftice, *Do to others as you would
have others do to you*, infpires all Men
with that other Maxim of natural Goodnefs a great deal lefs perfect, but perhaps
more ufeful, *Confult your own Happinefs
with as little Prejudice as you can to that
of others*. It is in a word, in this natural
Sentiment, rather than in fine-fpun Arguments, that we muft look for the Caufe of
that Reluctance which every Man would
experience to do Evil, even independently of the Maxims of Education. Tho'
it may be the peculiar Happinefs of *So-
crates* and other Geniufes of his Stamp,
to

to reafon themfelves into Virtue, the human Species would long ago have ceafed
to exift, had it depended entirely for its
Prefervation on the Reafonings of the
Individuals that compofe it.

With Paffions fo tame, and fo falutary
a curb, Men, rather wild than wicked,
and more attentive to guard againft Mif-
chief than to do any to other Animals,
were not expofed to any dangerous Dif-
fenfions: As they kept up no manner of
Correfpondence with each other, and were
of courfe Strangers to Vanity, to Refpect,
to Efteem, to Contempt; As they had no
Notion of what we call Meum and
Tuum, nor any true Idea of Juftice ; As
they confidered any Violence they were
liable to, as an Evil that could be eafily
repaired, and not as an Injury that de-
ferved Punifhment ; And as they never fo
much as dreamed of Revenge, unlefs perhaps mechanically and unpremeditatedly,

as

as a Dog who bites the Stone that has been thrown at him; their Diſputes could ſeldom be attended with bloodſhed, were they never occaſioned by a more conſiderable Stake than that of Subſiſtence: but there is a more dangerous Subject of Contention, which I muſt not leave unnoticed.

Among the Paſſions which ruffle the Heart of Man, there is one of a hot and impetuous Nature, which renders the Sexes neceſſary to each other; a terrible Paſſion which deſpiſes all Dangers, bears down all Obſtacles, and which in its Tranſports ſeems proper to deſtroy the human Species which it is deſtined to preſerve. What muſt become of Men abandoned to this lawleſs and brutal Rage, without Modeſty, without Shame, and every Day diſputing the Objects of their Paſſion at the Expence of their Blood?

We

We muſt in the firſt place allow that the more violent the Paſſions, the more neceſſary are Laws to reſtrain them: but beſides that the Diſorders and the Crimes, to which theſe Paſſions daily give riſe among us, ſufficiently prove the Inſufficiency of Laws for that Purpoſe, we would do well to look back a little further and examine, if theſe Evils did not ſpring up with the Laws themſelves; for at this Rate, tho' the Laws were capable of repreſſing theſe Evils, it is the leaſt that might be expected from them, ſeeing it is no more than ſtopping the Progreſs of a Miſchief which they themſelves have produced.

Let us begin by diſtinguiſhing between what is moral and what is phyſical in the Paſſion called Love. The phyſical Part of it is that general Deſire which prompts the Sexes to unite with each other; the moral Part is that which determines

termines this Defire, and fixes it upon
a particular Object to the Exclufion of
all others, or at leaft gives it a greater
Degree of Energy for this preferred Object. Now it is eafy to perceive that
the moral Part of Love is a factitious
Sentiment, engendered by Society, and
cried up by the Women with great Care
and Addrefs in order to eftablifh their
Empire, and fecure Command to that
Sex which ought to obey. This Sentiment, being founded on certain Notions
of Beauty and Merit which a Savage
is not capable of having, and upon Com-
parifons which he is not capable of making, can fcarcely exift in him: for as
his Mind was never in a Condition to
form abftract Ideas of Regularity and
Proportion, neither is his Heart fufceptible of Sentiments of Admiration and
Love, which, even without our perceiving it, are produced by our Application of thefe Ideas; he liftens folely to
the

the Difpofitions implanted in him by Nature, and not to Tafte which he never
was in a Way of acquiring; and every
Woman anfwers his Purpofe.

Confined entirely to what is phyfical
in Love, and happy enough not to know
thefe Preferences which fharpen the Appetite for it, at the fame Time that they
increafe the Difficulty of fatisfying fuch
Appetite, Men, in a State of Nature, muft
be fubject to fewer and lefs violent Fits of
that Paffion, and of courfe there muft be
fewer and lefs violent Difputes among them
in Confequence of it. The Imagination,
which caufes fo many Ravages among us,
never fpeaks to the Heart of Savages, who
peaceably wait for the Impulfes of Nature,
yield to thefe Impulfes without Choice and
with more Pleafure than Fury; and whofe
Defires never outlive their Neceffity for
the thing defired.

Nothing

Nothing therefore can be more evident, than that it is Society alone, which has added even to Love itself as well as to all the other Paffions, that impetuous Ardour, which fo often renders it fatal to Mankind; and it is fo much the more ridiculous to reprefent Savages conftantly murdering each other to glut their Brutality, as this Opinion is diametrically oppofite to Experience, and the Carribeans, the People in the World who have as yet deviated leaft from the State of Nature, are to all Intents and Purpofes the moft peaceable in their Amours, and the leaft fubject to Jealoufy, tho' they live in a burning Climate which feems always to add confiderably to the Activity of thefe Paffions.

As to the Inductions which may be drawn, in refpect to feveral Species of Animals, from the Battles of the Males, who in all Seafons cover our Poultry Yards

Yards with Blood, and in Spring particularly caufe our Forefts to ring again
with the Noife they make in difputing
their Females, we muft begin by exclu-
ding all thofe Species, where Nature has
evidently eftablifhed, in the relative Power
of the Sexes, Relations different from
thofe which exift among us: thus from
the Battles of Cocks we can form no
Induction that will affect the human Species. In the Species, where the Proportion is better obferved, thefe Battles muft
be owing entirely to the fewnefs of the
Females compared with the Males, or,
which is all one, to the exclufive Intervals,
during which the Females conftantly re-
fufe the Addreffes of the Males; for if the
Female admits the Male but two Months
in the Year, it is all the fame as if the
Number of Females were five fixths lefs
than what it is: now neither of thefe
Cafes is applicable to the human Species, where the Number of Females generally

nerally furpaffes that of Males, and where
it has never been obferved that, even
among Savages, the Females had, like
thofe of other Animals, ftated Times of
Paffion and Indifference. Befides, among
feveral of thefe Animals the whole Species takes Fire all at once, and for fome
Days nothing is to be feen among them
but Confufion, Tumult, Diforder and
Bloodfhed; a State unknown to the human Species where Love is never pe-
riodical. We cannot therefore conclude
from the Battles of certain Animals for
the Poffeffion of their Females, that the
fame would be the Cafe of Man in a
State of Nature; and tho' we might, as
thefe Contefts don't deftroy the other
Species, there is at leaft equal Room to
think they would not be fatal to ours;
nay it is very probable that they would
caufe fewer Ravages than they do in Society, efpecially in thofe Countries where,
Morality being as yet held in fome
Efteem, the Jealoufy of Lovers, and the
Ven.

Vengeance of Hufbands every Day produce Duels, Murders, and even worfe
Crimes; where the Duty of an eternal
Fidelity ferves only to propagate Adul-
tery; and the very Laws of Continence
and Honour neceffarily contribute to increafe Diffolutenefs, and multiply Abortions.

Let us conclude that favage Man, wandering about in the Forefts, without In-
duftry, without Speech, without any fixed
Refidence, an equal Stranger to War and
every focial Connection, without ftanding in
any fhape in need of his Fellows, as
well as without any Defire of hurting
them, and perhaps even without ever dif-
tinguifhing them individually one from
the other, fubject to few Paffions, and
finding in himfelf all he wants, let us, I fay,
conclude that favage Man thus circumftan-
ced had no Knowledge or Sentiment but
fuch as are proper to that Condition, that
he was alone fenfible of his real Neceffities,

ties, took notice of nothing but what it was his Intereſt to ſee, and that his Underſtanding made as little Progreſs as his Vanity. If he happened to make any Diſcovery, he could the leſs communicate it as he did not even know his Children. The Art periſhed with the Inventer; there was neither Education nor Improvement; Generations ſucceeded Generations to no Purpoſe; and as all conſtantly ſet out from the ſame Point, whole Centuries rolled on in the Rudeneſs and Barbarity of the firſt Age; the Species was grown old, while the Individual ſtill remained in a State of Childhood.

If I have enlarged ſo much upon the Suppoſition of this primitive Condition, it is becauſe I thought it my Duty, conſidering what ancient Errors and inveterate Prejudices I have to extirpate, to dig to the very Roots, and ſhew in a true
picture

picture of the State of Nature, how much even natural Inequality falls short in this State of that Reality and Influence which our Writers afcribe to it.

In Fact, we may eafily perceive that among the Differences, which diftinguifh Men, feveral pafs for natural, which are merely the Work of Habit and the different Kinds of Life adopted by Men living in a focial Way. Thus a robuft or delicate Conftitution, and the Strength and Weaknefs which depend on it, are oftener produced by the hardy or effeminate Manner in which a Man has been brought up, than by the primitive Conftitution of his Body. It is the fame thus in Regard to the Forces of the Mind; and Education not only produces a Difference between thofe Minds which are cultivated and thofe which are not, but even increafes that which is found among the firft in proportion

to

to their Culture; for let a Giant and a
Dwarf fet out in the fame Path, the Giant
at every Step will acquire a new Advantage over the Dwarf. Now, if we
compare the prodigious Variety in the
Education and Manner of living of the
different Orders of Men in a civil State,
with the Simplicity and Uniformity that
prevails in the animal and favage Life,
where all the Individuals make ufe of
the fame Aliments, live in the fame
Manner, and do exactly the fame things,
we fhall eafily conceive how much the
Difference between Man and Man in
the State of Nature muft be lefs than
in the State of Society, and how much
every Inequality of Inftitution muft increafe the natural Inequalities of the
Human Species.

But tho' Nature in the Diftribution
of her Gifts fhould really affect all the
Preferences that are afcribed to her, what
Ad-

Advantage could the moſt favoured derive from her Partiality, to the Prejudice of others, in a State of things, which
ſcarce admitted any kind of Relation
between her Pupils? Of what Service
can Beauty be, where there is no Love?
What will Wit avail People who don't
ſpeak, or Craft thoſe who have no Affairs to tranſact? Authors are conſtantly crying out, that the ſtrongeſt would
oppreſs the weakeſt; but let them explain what they mean by the Word Op-
preſſion. One Man will rule with Violence,
another will groan under a conſtant Subjection to all his Caprices : this is indeed preciſely what I obſerve among us,
but I don't ſee how it can be ſaid of ſa-
vage Men, into whoſe Heads it would be
a hard Matter to drive even the Meaning of the Words Domination and Servitude. One Man might indeed ſeize
on the Fruits which another had gathered,
on the Game which another had killed,

on

on the Cavern which another had occupied
for Shelter; but how is it possible he
should ever exact Obedience from him,
and what Chains of Dependence can there
be among Men who possess nothing? If I
am driven from one Tree, I have nothing
to do but look out for another; if one Place
is made uneasy to me, what can hinder me
from taking up my Quarters elsewhere?
But suppose I should meet a Man so
much superior to me in Strength, and
withal so wicked, so lazy and so barbarous as to oblige me to provide for
his Subsistence while he remains idle;
he must resolve not to take his Eyes
from me a single Moment, to bind me
fast before he can take the least Nap,
lest I should kill him or give him the
slip during his Sleep: that is to say, he
must expose himself voluntarily to much
greater Troubles than what he seeks to
avoid, than any he gives me. And after
all, let him abate ever so little of his
Vigilance; let him at some sudden Noise
but

59

but turn his Head another Way; I am already buried in the Foreſt, my Fetters are broke, and he never ſees me again.

But without inſiſting any longer upon theſe Details, every one muſt ſee that, as the Bonds of Servitude are formed merely by the mutual Dependence of Men one upon another and the reciprocal Neceſſities which unite them, it is impoſſible for one Man to enſlave another, without having firſt reduced him to a Condition in which he cannot live without the Enſlaver's Aſſiſtance; a Condition which, as it does not exiſt in a State of Nature, muſt leave every Man his own Maſter, and render the Law of the ſtrongeſt altogether vain and uſeleſs.

Having proved that the Inequality, which may ſubſiſt between Man and Man in a State of Nature, is almoſt imperceivable, and that it has very little Influence,

Influence, I muſt now proceed to ſhew
its Origin and trace its Progreſs, in the
ſucceſſive Developments of the human
Mind. After having ſhewed, that *Perfectibility*, the ſocial Virtues, and the other
Faculties, which natural Man had received in Potentia, could never be developed
of themſelves, that for that Purpoſe there
was a Neceſſity for the fortuitous Con-
currence of ſeveral foreign Cauſes, which
might never happen, and without which
he muſt have eternally remained in his
primitive Condition; I muſt proceed to
conſider and bring together the different
Accidents which may have perfected the
human Underſtanding by debaſing the
Species, render a Being wicked by rendering him ſociable, and from ſo remote a
Term bring Man at laſt and the World
to the Point in which we now ſee them.

I muſt own that, as the Events I am
about to deſcribe might have happened
many

many different Ways, my Choice of thefe
I fhall affign can be grounded on nothing but mere Conjecture; but befides
thefe Conjectures becoming Reafons, when
they are not only the moft probable that
can be drawn from the Nature of things;
but the only Means we can have of dif-
covering Truth, the Confequences I mean
to deduce from mine will not be merely
conjectural, fince, on the Principles I have
juft eftablifhed, it is impoffible to form
any other Syftem, that would not fupply
me with the fame Refults, and from
which I might not draw the fame Con-
clufions.

6

Reprinted from pages 183–184 of *Sir John Chardin's Travels in Persia*, Argonaut Press, London, 1927 (originally published in 1720 by J. Smith)

OF THE TEMPER, MANNERS, AND CUSTOMS OF THE PERSIANS

J. Chardin

THE *Persian* Blood is naturally thick; it may be seen by the *Guebres*, who are the remainder of the ancient *Persians*; they are homely, ill shap'd, dull, and have a rough Skin, and an Olive Complexion. The same Thing is observ'd also in the Provinces next the *Indus*, whereof the Inhabitants are little better shap'd than the *Guebres*, because they marry only amongst them: But in the other Parts of the Kingdom, the *Persian* Blood is now grown clearer, by the mixture of the *Georgian* and *Circassian* Blood, which is certainly the People of the World, which Nature favours most, both upon the Account of the Shape and Complexion, and of the *Boldness* and *Courage*; they are likewise *Sprightly*, *Courtly* and *Amorous*. There is scarce a Gentleman in *Persia*, whose Mother is not a *Georgian*, or a *Circassian* Woman; to begin with the King, who commonly is a *Georgian*, or a *Circassian* by the Mother's side; and whereas, that Mixture begun above a hundred Years ago, the Female kind is grown fairer, as well as the other, and the *Persian* Women are now very *handsome*, and very well *shap'd*, tho' they are still inferior to the *Georgians*: As to the Men, they are commonly *Tall*, *Straight*, *Ruddy*, *Vigorous*, have a good Air, and a pleasant Countenance. The Temperateness of their Climate, and the Temperance they are brought up in, do not a little contribute to their Shape and Beauty. Had it not been for the Alliance before mention'd, the Nobility of *Persia* had been the ugliest Men in the World; for they originally came from those Countries between *China* and the *Caspian* Sea, call'd *Tartary*; the Inhabitants whereof being the homeliest Men of *Asia*, are short and thick, have their Eyes and Nose like the *Chinese*, their Face flat and broad, and their Complexion yellow, mix'd with black.

[*Editor's Note:* Material has been omitted at this point.]

7

Reprinted from pages 62–77 of *Essay on the Causes of the Variety of Complexion and Figure in the Human Species,* R. Aitken, Philadelphia, 1787

ESSAY ON THE CAUSES OF THE VARIETY OF COMPLEXION AND FIGURE IN THE HUMAN SPECIES

S. Smith

[*Editor's Note:* In the original, material precedes and follows this excerpt.]

The ſtate of ſociety comprehends diet, clothing, lodging, manners, habits, face of the

country, objects of science, religion, interests, passions and ideas of all kinds, infinite in number and variety. If each of these causes be admitted to make, as undoubtedly they do, a small variation on the human countenance, the different combinations and results of the whole must necessarily be very great; and combined with the effects of climate will be adequate to account for all the varieties we find among mankind*.

Another origin of the varieties springing from the state of society is found in the power which men possess over themselves of producing great changes in the human form, according to any common standard of beauty which they may have adopted. The standard of human beauty, in any country, is a general idea formed from the combined effect of climate and of the state of society. And it

* As all these principles may be made to operate in very different ways, the effect of one may, often, be counteracted, in a degree, by that of another. And climate will essentially change the effects of all. The people in different parts of the same country may, from various combinations of these causes, be very different. And, from the variety of combination, the poor of one country may have better complexion, features and proportions of person, than those in another, who enjoy the most favourable advantages of fortune. Without attention to these circumstances, a hasty observer will be apt to pronounce the remarks in the essay to be ill-founded, if he examines the human form in any country by the effect that is said to arise from one principle alone, and do not, at the same time, take in the concomitant or correcting influence of other causes.

it reciprocally contributes to increafe the effect from which it fprings.   Every nation
varies as much from others in ideas of beauty as in perfonal appearance.   Whatever be
that ftandard, there is a general effort to at-
tain it, with more or lefs ardor and fuccefs,
in proportion to the advantages which men
poffefs in fociety, and to the eftimation in
which beauty is held.

To this object tend the infinite pains to
compofe the features, and to form the attitudes of children, to give them the gay and
agreeable countenance that is created in company, and to exftinguifh all deforming emotions of the paffions.   To this object tend
many of the arts of polifhed life.   How
many drugs are fold, and how many applications are made for the improvement of beauty? how many artifts of different kinds live
upon this idea of beauty? If we dance, beauty is the object; if we ufe the fword, it is
more for beauty than defence.   If this general effort after appearance fometimes leads
the decrepid and deformed into abfurdity, it
has, however, a great and national effect.——
Of its effect in creating diftinctions among
nations in which different ideas prevail and
different

different means are employed for attaining them, we may frame fome conception, from the diftinctions that exift in the fame nation, in which fimilar ideas and fimilar means are ufed, only in different degrees. What a difference is there between the foft and elegant tints of complexion in genteel life, and the coarfe ruddinefs of the vulgar?—between the uncouth features and unpliant limbs of an unpolifhed ruftic, and the complacency of countenance, the graceful and eafy air and figure of an improved citizen?—between the fhaped and meaning face of a well bred lady, and the foft and plump fimplicity of a country girl?—We now eafily account for thefe differences, becaufe they are familiar to us, or, becaufe we fee the operation of the caufes. But if we fhould find an intire nation diftinguifhed by one of thefe characters, and another by the contrary, fome writers would pronounce them different races; although a true philofopher ought to underftand that the cultivation of oppofite ideas of beauty muft have a greater effect in diverfifying the human countenance, than various degrees, or modes, of cultivating the fame ideas. The countenance of Europe was more various, three centuries ago, than it is at prefent. The

diverfities

diverfities that depend upon this caufe are infenfibly wearing away as the progrefs of refinement is gradually approximating the manners and ideas of the people to one ftandard. But the influence of a general idea, or ftandard, of the human form; and the pains taken, or the means employed, to bring our own perfons to it, are through their familiarity often little obferved. The means employed by other nations, who aim at a different idea, attract more notice by their novelty.—The nations beyond the Indus, as well as the Tartars, from whom they feem to have derived their ideas of beauty with their origin*, univerfally admire fmall eyes and large ears. They are at great pains, therefore, to comprefs their eyes at the corners, and to ftretch their ears by heavy weights appended to them, by drawing them frequently with the hand, and by cutting their rims, fo that they may hang down to their fhoulders, which they confider as the higheft beauty. On the fame principle, they extirpate the hair from their bodies;

---

* It is probable that the countries of India and China might have been peopled before the regions of Tartary; but, the frequent conquefts which they have fuffered, and particularly the former, from Tartarian nations, have changed their habits, ideas and perfons, even more perhaps than Europe was changed by the deluge of barbarians that overwhelmed it in the fifth century. The prefent nations beyond the Indus are, in effect, Tartars changed by the power of climate, and of a new ftate of fociety.

bodies; and, on the face, they leave only a few tufts here and there which they fhave. The Tartars often extirpate the whole hair of the head, except a knot on the crown, which they braid and adorn in different manners. Similar ideas of beauty with regard to the eyes, the ears and the hair; and fimilar cuftoms, in the Aborigines of America, are no inconfiderable proofs that this continent has been peopled from the north-eaftern regions of Afia*. In Arabia and Greece large eyes are efteemed beautiful; and in thefe countries they take extraordinary pains to ftretch the lids, and extend their aperture. In India, they dilate the forehead in infancy, by the application of broad plates of lead. In China they comprefs the feet. In Caffraria, and many other parts of Africa, and in Lapland, they

---

* The celebrated Dr. Robertfon, in his hiftory of America, deceived by the mifinformation of hafty or ignorant obfervers, has ventured to affert that the natives of America have no hair on their face or on their body; and like many other philofophers, has fet himfelf to account for a fact that never exifted. It may be laid down almoft as a general maxim, that the firft relations of travellers are falfe. They judge of appearances in a new country under the prejudices of ideas and habits contracted in their own. They judge from particular inftances, that may happen to have occurred to them, of the ftature, the figure and the features of a whole nation. Philofophers ought never to admit a fact on the relations of travellers, till their characters for intelligence and accurate obfervation be well afcertained; nor even then, till the obfervation has been repeated, extended, and compared in many different lights, with other facts. The Indians have hair on the face and body; but from a falfe fenfe of beauty they extirpate it with great pains. And traders among them are well informed, that tweezers for that purpofe, are profitable articles of commerce.

they flatten the nofe in order to accomplifh a
capricious idea of beauty. The fkin, in many
nations is darkened by art; and all favages
efteem certain kinds of deformity to be perfections; and ftrive to heighten the admiration of their perfons, by augmenting the wild-
nefs of their features. Through every country on the globe we might proceed in this
manner, pointing out the many arts which the
inhabitants practife to reach fome favourite
idea of the human form. Arts that infenfibly, through a courfe of time, produce a great
and confpicuous effect. Arts which are ufually fuppofed to have only a perfonal influence; but which really have an operation on
pofterity alfo. The procefs of nature in this
is as little known as in all her other works.
The effect is frequently feen. Every remarkable change of feature that has grown into a
habit of the body, is tranfmitted with other
perfonal properties, to offspring. The coarfe
features of labouring people, created by hard-
fhips, and by long expofure to the weather,
are communicated.—The broad feet of the
ruftic, that have been fpread by often treading
the naked ground; and the large hand and
arm, formed by conftant labour, are difcern-
ible in children. The increafe or diminution
of

of any other limb or feature formed by habits that aim at an idea of beauty, may, in like manner, be imparted. We continually fee the effect of this principle on the inferior animals. The figure, the colour and properties of the horfe are eafily changed according to the reigning tafte. Out of the fame original ftock the Germans who are fettled in Pennfylvania, raife large and heavy horfes; the Irifh raife fuch as are much lighter and fmaller. According to the pains beftowed, you may raife from the fame race, horfes for the faddle and horfes for the draught. Even the colour can be fpeedily changed according as fafhion is pleafed to vary its caprice. And, if tafte prefcribes it, the fineft horfes fhall, in a fhort time, be black, or white, or bay*. Human nature much more pliant, and affected by a greater variety of caufes from food, from clothing, from lodging and from manners, is ftill more eafily fufceptible of change, according to any general ftandard, or idea of the human form. To this principle, as well as to the manner of living, it may be, in part, attributed that the Germans, the Swedes and the French, in different parts of the United States, who live chiefly among themfelves, and

---

* By chufing horfes of the requifite qualities, to fupply the ftuds.

and cultivate the habits and ideas of the countries from which they emigrated, retain, even
in our climate, a ftrong refemblance of their
primitive ftocks.   Thofe, on the other hand,
who have not confined themfelves to the contracted circle of their countrymen, but have
mingled freely with the Anglo-Americans,
entered into their manners, and adopted their
ideas, have affumed fuch a likenefs to them,
that it is not eafy now to diftinguifh from one
another people who have fprung from fuch
different origins.

I have faid that the procefs of nature in
this, as in all her other works, is inexplicable.   One fecondary caufe, however, may be
pointed out, which, feems to have confidera-
ble influence on the event*.   Connexions
in marriage will generally be formed on this
idea of human beauty in any country.   An
influence

---

* Befides this, men will foon difcover thofe kinds of diet, and thofe
modes of living that will be moft favourable to their ideas. The power
of imagination in pregnant women, might perhaps deferve fome confider-
ation on this fubject.   Some years fince, this principle was carried to ex-
cefs. I am ready to believe that philofophers, at prefent, run to extremes
on the other hand. They deny intirely the influence of imagination.
But fince the emotions of fociety have fo great an influence, as it is evident they have, in forming the countenance; and fince the refemblance
of parents is communicated to children, why fhould it be deemed incredible that thofe general ideas which contribute to form the features of the
parent, fhould contribute alfo to form the features of the child.

influence this which will gradually approximate the countenance towards one common standard. If men in the affair of marriage, were as much under management as some other animals, an absolute ruler might accomplish, in his dominions almost any idea of the human form. But, left as this connexion is to the passions and interests of individuals, it is more irregular and imperfect in its operations. And the negligence of the vulgar, arising from their want of taste, impedes, in some degree, the general effect. There is however a common idea which men insensibly to themselves, and almost without design, pursue. And they pursue it with more or less success in proportion to the rank and taste of different classes in society, where they do not happen in particular instances, to be governed in connexions of marriage by interest ever void of taste. The superior ranks will always be first, and, in general, most improved, according to the prevalent idea of national beauty; because, they have, it more than others, in their power to form matrimonial connexions favourable to this end. The Persian nobility, improved in their idea of beauty, by their removal to a new climate, and a new state of society, have, within a few races,

almost

almoſt effaced the characters of their Tartarian origin. The Tartars, from whom they are
deſcended, are among the moſt deformed and
ſtupid nations upon earth. The Perſians by
obtaining the moſt beautiful and agreeable
women from every country, are become a
tall, and well featured, and ingenious nation.
The preſent nations of Europe have with the
refinement of their manners and ideas changed
and refined their perſons. Nothing can exceed the pictures of barbariſm and deformity
given us of their anceſtors, by the Roman
writers. Nothing can exceed the beauty of
many of the preſent women of Europe and
America who are deſcended from them.
And the Europeans, and Americans are, the
moſt beautiful people in the world, chiefly,
becauſe their ſtate of ſociety is the moſt improved. Such examples tend to ſhew how
much the varieties of nations may depend on
ideas created by climate, adopted by inheritance, or formed by the infinite changes of
ſociety and manners*. They ſhew, likewiſe
how

---

* Society in America is gradually advancing in refinement: and if my
obſervation has been juſt the preſent race furniſhes more women of exqui-
ſite beauty than the laſt, though they may not always be found in the
ſame families. And if ſociety ſhould continue its progreſſive improvement, the next race may furniſh more than the preſent. Europe has certainly made great advances in refinement of ſociety, and probably in beauty.

how much the human race might be improved both in perſonal and in mental qualities,
by a well-directed care.

The ancient Greeks ſeem to have been the
people moſt ſenſible of its influence. Their
cuſtoms, their exerciſes, their laws, and their
philoſophy, appear to have had in view, among
other objects, the beauty and vigour of the
human conſtitution. And it is not an improbable conjecture, that the fine models exhibited, in that country, to ſtatuaries and
painters, were one cauſe of the high perfection to which the arts of ſculpture and painting arrived in Greece. If ſuch great improvements were introduced by art into the human
figure, among this elegant and ingenious people, it is a proof at once of the influence of
general ideas, and of how much might be effected by purſuing a juſt ſyſtem upon this
ſubject. Hitherto, it has been abandoned too
much to the government of chance. The
great and noble have uſually had it more in
their power than others to ſelect the beauty
of nations in marriage: and thus, while,

without

ty. And if exact pictures could have been preſerved of the human countenance and form in every age ſince the great revolution made by the
barbarians, we ſhould, perhaps, find Europe as much improved in its features as in its manners.

without fyftem or defign, they gratified only their own tafte, they have generally diftinguifhed their order, as much by elegant proportions of perfon, and beautiful features, as by its prerogatives in fociety. And the tales of romances that defcribe the fuperlative beauty of captive princeffes, and the fictions of poets, who characterife their kings and nobles, by uncommon dignity of carriage and elegance of perfon, and by an elevated turn of thinking, are not to be afcribed folely to the venality of writers prone to flatter the great, but have a real foundation in nature*. The ordinary ftrain of language, which is borrowed from nature, vindicates this criticifm. A *princely* perfon, and a *noble* thought, are ufual figures of fpeech†.—Mental capacity, which is as various as climate, and as

perfonal

---

* Coincident with the preceding remarks on the nations of Europe, is an obfervation made by Capt. Cook, in his laft voyage, on the ifland Ohwyhee, and oh the iflands in general, which he vifited in the great fouth fea. He fays, " the fame fuperiority which is obfervable in the " *Erees* [or nobles] through all the other iflands, is found alfo here. Thofe " whom we faw, were, without exception perfectly well formed ; where- " as the lower fort, befides their general inferiority, are fubject to all the " variety of make and figure that is feen in the *populace* of other countries." Cook's third voyage, book 3d, chap. 6th.

† Such is the deference paid to beauty, and the idea of fuperiority it infpires, that to this quality, perhaps, does the body of princes and nobles, collectively taken, in any country, owe great part of their influence over the populace. Riches and magnificence in drefs and equipage, produce

much

perfonal appearance, is, equally with the latter, fufceptible of improvement, from fimilar
caufes. The body and mind have fuch mutual influence, that whatever contributes to
change the human conftitution in its form or
afpect, has an equal influence on its powers
of reafon and genius. And thefe have again
a reciprocal effect in forming the countenance.
One nation may, in confequence of conftitu-
tional peculiarities, created more, perhaps, by
the ftate of fociety, than by the climate, be
addicted to a grave and thoughtful philofo-
phy; another may poffefs a brilliant and
creative imagination; one may be endowed
with acutenefs and wit; another may be diftinguifhed for being phlegmatic and dull.
Bæotian and *Attic* wit was not a fanciful, but
real diftinction, though the remote origin of
Cadmus and of Cecrops was the fame. The
ftate of manners and fociety in thofe repub-
lics produced this difference more than the
Bœotion air, to which it has been fo often attributed. By the alteration of a few political,
or civil, or commercial inftitutions, and con-
fequently,

much of their effect by giving an artificial beauty to the perfon. How
often does hiftory remark that young princes have attached their fubjects,
and generals their foldiers, by extraordinary beauty? And young and
beautiful queens have ever been followed and ferved with uncommon en-
thufiafm.

sequently, of the objects of society and the train of life, the establishment of which depended on a thousand accidental causes, Thebes might have become Athens, and Athens Thebes. Different periods of society, different manners, and different objects, unfold and cultivate different powers of the mind. Poetry, eloquence, and philosophy seldom flourish together in their highest lustre. They are brought to perfection by various combinations of circumstances, and are found to succeed one another in the same nation at various periods, not because the race of men, but because manners and objects are changed. If as faithful a picture could be left to posterity of personal as of mental qualities, we should probably find the one, in these several periods, as various as the other; and we should derive from them a new proof of the power of society to multiply the varieties of the human species. Not only deficiency of objects to give scope to the exercise of the human intellect is unfavourable to its improvement; but all rudeness of manners is unfriendly to the culture, and the existence of taste, and even coarse and meagre food may have some tendency to blunt the powers of genius. These causes have a more powerful

operation

operation than has hitherto been attributed to them by philofophers; and merit a more minute and extenfive illuftration than the fubject of this difcourfe will admit. The mental capacities of favages, for thefe caufes, are ufually weaker than the capacities of men in civilized fociety*. The powers of their minds, through defect of objects to employ them, lie dormant, and even become extinct. The faculties which, on fome occafions, they are found to poffefs, grow feeble through want of motives to call forth their exercife. The coarfenefs of their food, and the filthinefs of their manners tend to blunt their genius. And the Hottentots, the Laplanders, and the people of New-Holland are the moft ftupid of mankind for this, among other reafons, that they approach, in thefe refpects, the neareft to the brute creation†.

---

* The exaggerated reprefentations which we fometimes receive of the ingenuity and profound wifdom of favages, are the fruits of weak and ignorant furprize. And favages are praifed by fome writers for the fame reafon that a monkey is—a certain imitation of the actions of men in fociety, which was not expected from the rudenefs of their condition. There are doubtlefs degrees of genius among favages as well as among civilized nations; but the comparifon fhould be made of favages among themfelves; and not of the genius of a favage, with that of a polifhed people.

† It is well known that the Africans who have been brought to America, are daily becoming, under all the difadvantages of fervitude, more ingenious and fufceptible of inftruction. This effect, which has been taken notice of more than once, may, in part perhaps, be attributed to a change in their modes of living, as well as to fociety, or climate.

8

Reprinted from pages 502–505 of *Zoonomia; Or The Laws of Organic Life,*
vol. 1, Johnson, London, 1794, 586p.

OF GENERATION

E. Darwin

[*Editor's Note:* In the original, material precedes and follows this excerpt.]

When we confider all thefe changes of animal form, and innumerable others, which may be collected from the books of natural hiftory; we cannot but be convinced, that the fetus or embryon is formed by appofition of new parts, and not by the diftention of a primordial neft of germs, included one within another, like the cups of a conjurer.

Fourthly, when we revolve in our minds the great fimilarity of ftructure, which obtains in all the warm-blooded animals, as well quadrupeds, birds, and amphibious animals, as in mankind; from the moufe and bat to the elephant and whale; one is led to conclude, that they have alike been produced from a fimilar living filament. In fome this filament in its advance to maturity has acquired hands and fingers, with a fine fenfe of touch, as in mankind. In others it has acquired claws or talons, as in tygers and eagles. In others, toes with an intervening web, or membrane, as in feals and geefe. In others it has acquired cloven hoofs, as in cows and fwine; and whole hoofs in others, as in the horfe. While in the bird kind this original living filament has put forth wings inftead of arms or legs, and feathers inftead of hair. In fome it has protruded horns on the forehead inftead of teeth in the fore part of the upper jaw; in others tufhes inftead of horns; and in others beaks inftead of either. And all this exactly as is daily feen in the tranfmutations of the tadpole, which acquires legs and lungs, when he wants them; and lofes his tail, when it is no longer of fervice to him.

Fifthly, from their firft rudiment, or primordium, to the termination of their lives, all animals undergo perpetual transformations;
which

80

which are in part produced by their own exertions in confequence of their defires and averfions, of their pleafures and their pains, or of irritations, or of affociations; and many of thefe acquired forms or propenfities are tranfmitted to their pofterity. See Sect. XXXI. 1.

As air and water are fupplied to animals in fufficient profufion, the three great objects of defire, which have changed the forms of many animals by their exertions to gratify them, are thofe of luft, hunger, and fecurity. A great want of one part of the animal world has confifted in the defire of the exclufive poffeffion of the females; and thefe have acquired weapons to combat each other for this purpofe, as the very thick, fhield-like, horny fkin on the fhoulder of the boar is a defence only againft animals of his own fpecies, who ftrike obliquely upwards, nor are his tufhes for other purpofes, except to defend himfelf, as he is not naturally a carnivorous animal. So the horns of the ftag are fharp to offend his adverfary, but are branched for the purpofe of parrying or receiving the thrufts of horns fimilar to his own, and have therefore been formed for the purpofe of combating other ftags for the exclufive poffeffion of the females; who are obferved, like the ladies in the times of chivalry, to attend the car of the victor.

The birds, which do not carry food to their young, and do not therefore marry, are armed with fpurs for the purpofe of fighting for the exclufive poffeffion of the females, as cocks and quails. It is certain that thefe weapons are not provided for their defence againft other adverfaries, becaufe the females of thefe fpecies are without this armour. The final caufe of this conteft amongft the males feems to be, that the ftrongeft and moft active animal fhould propagate the fpecies, which fhould thence become improved.

Another great want confifts in the means of procuring food, which has diverfified the forms of all fpecies of animals. Thus the nofe of the fwine has become hard for the purpofe of turning up the foil in

fearch

search of insects and of roots. The trunk of the elephant is an elongation of the nose for the purpose of pulling down the branches of trees
for his food, and for taking up water without bending his knees.
Beasts of prey have acquired strong jaws or talons. Cattle have acquired a rough tongue and a rough palate to pull off the blades of grass,
as cows and sheep. Some birds have acquired harder beaks to crack
nuts, as the parrot. Others have acquired beaks adapted to break the
harder seeds, as sparrows. Others for the softer seeds of flowers, or
the buds of trees, as the finches. Other birds have acquired long
beaks to penetrate the moister soils in search of insects or roots, as
woodcocks; and others broad ones to filtrate the water of lakes, and
to retain aquatic insects. All which seem to have been gradually produced during many generations by the perpetual endeavour of the creatures to supply the want of food, and to have been delivered to their
posterity with constant improvement of them for the purposes required.

The third great want amongst animals is that of security, which
seems much to have diversified the forms of their bodies and the colour
of them; these consist in the means of escaping other animals more
powerful than themselves. Hence some animals have acquired wings
instead of legs, as the smaller birds, for the purpose of escape. Others
great length of fin, or of membrane, as the flying fish, and the bat.
Others great swiftness of foot, as the hare. Others have acquired
hard or armed shells, as the tortoise and the echinus marinus.

The contrivances for the purposes of security extend even to vegetables, as is seen in the wonderful and various means of their concealing or defending their honey from insects, and their seeds from
birds. On the other hand swiftness of wing has been acquired by
hawks and swallows to pursue their prey; and a proboscis of admirable
structure has been acquired by the bee, the moth, and the humming
bird, for the purpose of plundering the nectaries of flowers. All
which

which feem to have been formed by the original living filament, excited into action by the neceffities of the creatures, which poffefs them, and on which their exiftence depends.

From thus meditating on the great fimilarity of the ftructure of the warm-blooded animals, and at the fame time of the great changes they undergo both before and after their nativity; and by confidering in how minute a portion of time many of the changes of animals above defcribed have been produced; would it be too bold to imagine, that in the great length of time, fince the earth began to exift, perhaps millions of ages before the commencement of the hiftory of mankind, would it be too bold to imagine, that all warm-blooded animals have arifen from one living filament, which THE GREAT FIRST CAUSE endued with animality, with the power of acquiring new parts, attended with new propenfities, directed by irritations, fenfations, volitions, and affociations; and thus poffeffing the faculty of continuing to improve by its own inherent activity, and of delivering down thofe improvements by generation to its pofterity, world without end!

9

Reprinted from pages 15-16 of *The Art of Improving the Breeds of Domestic Animals. In a Letter Addressed to the Right Hon. Sir Joseph Banks, K.B.,* J. Harding, London, 1809

THE ART OF IMPROVING THE BREEDS OF DOMESTIC ANIMALS. IN A LETTER ADDRESSED TO THE RIGHT HON. SIR JOSEPH BANKS, K.B.

J. Sebright

[*Editor's Note:* In the original, material precedes and follows this excerpt.]

Many causes combine to prevent animals, in a state of nature, from degenerating; they are perpetually intermixing, and therefore do not feel the bad effects of breeding *in-and-in*: the perfections of some correct the imperfections of others, and they go on without any material alteration, except what arises from the effects of food and climate.

The greatest number of females will, of course, fall to the share of the most vigorous males; and the strongest individuals of both sexes, by driving away the weakest, will enjoy the best food, and the most favourable situations, for themselves and for their offspring.

A severe winter, or a scarcity of food, by destroying the weak and the unhealthy, has all the good effects of the most skilful selection. In cold and barren countries no animals can live to the age of maturity, but those who have strong constitutions; the weak and the unhealthy do not live to propagate their infirmities, as is too often the case with our domestic animals. To this I attribute the peculiar hardiness of the horses, cattle, and sheep, bred in mountainous countries, more than to their having been inured to the severity of the climate; for our domestic animals do not become more hardy by being exposed, when young, to cold and hunger: animals so treated will not, when arrived at the age of maturity, endure so much hardship as those who have been better kept in their infant state.

10

Reprinted from pages 37–46 of *Researches into the Physical History of Man*, J. and A. Arch, London, 1813

ON DIVERSITIES OF FORM

J. Prichard

[*Editor's Note:* In the original, material precedes and follows this excerpt.]

(*a*) It is said that in every different state or province of Italy, the people have their peculiar form of features, or characteristic physiognomy. This fact must be accounted for on the principle above stated, for no other cause can be imagined.

The different casts of people in Hindustan, who are settled in the same country, or who wander over it, have been prevented by the strict prohibitions of their religion from intermarriages with each other for many ages. The result of this long continued experiment is illustrative of the foregoing remarks. Each of these casts has acquired, though all of them are subject to the same local causes, a distinct set of features, and they are all easily known by people who are conversant with them.(*b*)

From similar causes, the difference of features which we remark between the English and Scottish people, and between the French and Italians, must be supposed to have arisen. We cannot imagine diversity of origin, or any considerable effect arising from difference of soil and climate in either of these instances. And perhaps the distinct physiognomy which characterizes the several nations of Europe may be in great part accounted for on the same principles.

The hereditary tendency of peculiar corporeal structure in the brute species has long been

(*a*) This fact is asserted by all travellers in Italy.
(*b*) Major Orme's Indostan. Introduction.

matter of common observation, and it is on the skilful application of it, that the art of the breeders of cattle, horses, and other domesticated animals consists. The power which human art possesses of modifying the individual, is very limited indeed; but by diligently taking advantage of the natural tendency to transmit any qualities which happen to arise, a very considerable influence is exercised over the race. Different breeds are thus formed endowed with divers properties, which render them useful in various ways to their owners. The process consists in a careful selection of those individual animals, which happen to be possessed in a more remarkable degree than the generality, of the characters which it is desirable to perpetuate. These are kept for the future propagation of the stock, and a repeated attention is paid to the same circumstances, till the effect continually increasing, a particular figure, colour, proportion of limbs or any other attainable quality, is established in the race, and the conformity is afterwards maintained by removing from the breed any new variety which may casually spring up in it.

Thus it has long been a favourite caprice among the farmers of different counties of England, to encourage breeds of cattle of peculiar colours. In some counties they have chosen to have all their stock of oxen brown; in others they have them spotted in a particular manner. In such places varieties thus rendered general, become to a great degree constant, and animals of a different character from that of the race in this man ner constituted, are very rarely produced.

It is perhaps to a similar diversity of choice in the breeders, that we find in some districts of our country, sheep and oxen, of which the whole breeds are horned. In other places they are altogether destitute of horns.

These instances are of an inferior class, though they exemplify the general principle; but it is capable of a much more useful application. By the same process distinct breeds of animals, as of horses for example, are formed, which are adapted by their peculiar conformation to various purposes of utility. Strength and the more unwieldy form, necessary to great power of limbs, become the character of one race of horses, while another is distinguished for a light and more graceful shape, favourable to agility and celerity of motion. The finer breeds of horses have perhaps attained greater elegance and perfection in England, than was ever to be found in the species in any other country, and this is to be attributed to the great attention which has been bestowed on their propagation, owing to the prevalent fashion of horse-racing.(a) We find from the accounts of Cæsar and Tacitus,(b) that

(a) Cæsar de Bello Gallico.
(b) Tacitus de Mor. Germanorum.

the horses of Germany were formerly much inferior to those of Gaul. But the German breeds
have in the present day greatly the advantage of
the French. The change must be ascribed to
the more careful and scientific management of
the propagators in the former country.

Perhaps it has arisen from the same care in
the formation of breeds that we find among the
varieties of Dogs, one race remarkable for acute
sight, another for fine scent, and a third of which
the greater strength and weight of limbs, point
them out as fit for the purpose of nightly protection. The instinct varies in all these instances,
as we might expect from analogy, with the peculiarities of organization. This principle seems
in general to direct every animal to seek its sub-
sistence, in the way for which its corporeal
structure happens best to qualify it. Accordingly we find considerable diversities of instinct
within the limits of the same species.

If the same constraint were exercised over
men, which produces such remarkable effects
among the brute kinds, there is no doubt that
its influence would be as great. But no despot
has ever thought of amusing himself in this
manner, or at least such an experiment has never
been carried on upon that extensive scale, which
might lead to important results.(*a*) Certain moral

(*a*) Something of this kind was indeed attempted by the
kings of Prussia, but their project referred to stature.

causes however, have an influence on mankind, which appears in some degree to lead to similar ends.

(*a*) The perception of beauty is the chief principle in every country which directs men in their marriages. It does not appear that the inferior tribes of animals have any thing analogous to this feeling, but in the human kind it is universally implanted. It is very obvious that this peculiarity in the constitution of man, must have considerable effects on the physical character of the race, and that it must act as a constant principle of improvement, supplying the place in our own kind of the beneficial controul which we exercise over the brute creation. This is probably the final cause for which the instinctive perception of human beauty was implanted by Providence in our nature. For the idea of beauty of person, is synonymous with that of health and perfect organization.

In the ruder stages of society the natural principles operate with more undisturbed energy. In all nations that have not attained a high degree of civilization and refinement, we find beauty to be the only qualification in the female, to which the least value or importance is affixed.

(*a*) D. S. S. Smith of New Jersey in America, in an essay on the causes of variety in the figure and complexion of the human species, has made some ingenious remarks on this subject.

The effect therefore of this principle must be much greater, and more conspicuous in barbarous communities, than among civilized people, but it is every where on the great scale of considerable moment.

The disgust, which instances of deformity naturally excite, prevents the hereditary transmission of such peculiarities, which would probably in many cases happen, if deformed persons were generally married. The greater examples of malconformation would be frequently found to be conjoined with sterility, but that is not the case with lesser instances; and these might be rendered general and perpetual, if that evil were not guarded against by a provision of instinct. Among savage tribes, the repugnance felt at the view of any deformed appearance in the human kind is so strong, that it is said to be the general custom in such nations, to destroy children which are imperfect in their figure. The same practice prevailed among the Lacedemonians, and several other nations of antiquity.

In countries where the people are divided into different ranks or orders of society, which is almost universally the case, the improvement of person, which is the result of the abovementioned cause, will always be much more conspicuous in the higher than in the inferior classes. The former are guided in their marriages, as in all the other actions of life, by their inclinations. The

latter are governed, especially where servile subjection is established, by the caprice of their
superiors, or by motives of convenience or necessity. The noble families of modern Persia
were originally descended from a tribe of ugly
and bald-headed Mongoles. They have constantly
selected for their harams the most beautiful
females of Circassia. The race has been thus
gradually ameliorated, and is said now to exhibit fine and comely persons. In states thus
situated with respect to their political circumstances, the inferior people must in many instances suffer deterioration, while the higher
rank improves. This must constitute a very
marked difference in the aspect of the two
orders.

Such diversity is every where observed. It
has been remarked, where we should scarcely
have looked for it, viz. among the barbarous
islanders of the Pacific. Capt. Cook, in describing the people of Owhyhee, says " The same
" superiority which is observed in the Erees
" (nobles) in all other islands is found also here.
" Those whom we saw were without exception
" perfectly well formed, whereas the lower sort,
" besides their general inferiority, are subject to
" all the variety of make and figure that is seen
" in the populace of other countries." *(a)*

(a) Cook's last voyage. **Book 3.**

The same observation is equally applicable to the inhabitants of most of the European countries.

Since it appears that the prevalent idea of beauty acting as a constant principle upon one nation during a long time produces a remarkable effect, it is to be supposed that if different standards prevailed in several countries, their influence would tend to establish a considerable diversity. It is probable that the natural idea of the beautiful in the human person has been more or less distorted in almost every nation. Peculiar characters of countenance, in many countries accidentally enter into the ideal standard. This observation has been made particularly of the Negroes of Africa, who are said to consider a flat nose and thick lips as principal ingredients of beauty, and we are informed by Pallas that the Kalmucs *(a)* esteem no face as handsome, which has not the eyes in angular position, and the other characteristics of their race.*(b)* The Aztecs of Mexico have ever preferred a depressed forehead, which forms the strongest contrast to the majestic contour of the Grecian busts; the former represented their divinities with a head more flattened than it is ever seen among the

(a) Pallas. Voyages en Siberie. French traslation.
(b) Humboldt's political essay on the kingdom of New Spain. Vol. I.

Caribs, and the Greeks on the contrary gave to their gods and heroes a still more unnatural elevation. We do not attempt thus to account for all the peculiarities of these races, for the variety in the opinion of beauty, may be in some part the effect as well as the cause of national diversity, but we adduce these instances to exemplify a principle, the effects of which must, as we conceive, be very important, and tend to widen, if they have not in the first instance produced, the physical differences of nations.

These remarks were so obviously connected with the observations made in a former page, on the disposition manifested by all living species to assume varieties of figure, and on the tendency which such varieties in the animal kingdom evince to become permanent in the race, that we have ventured to follow them out, though they have led us to digress in some measure from the order of our argument. They will be useful in our inquiry concerning the nature of national diversities, and may enable us to explain some peculiar appearances. But they do not afford any direct solution of the question now before us, which is whether the differences in form, that are found to subsist between the European, the Ethiopian, and the Mongole, or in general whether the greatest examples of such diversity, which are observed in mankind, are specific differences, or only instances of deviation. In order to solve this doubt, we must adopt a more systematical method of inquiry, with reference to the analogical reasoning proposed in the foregoing pages.

Part II

THE DEVELOPMENT OF THE THEORY
OF SEXUAL SELECTION BY
CHARLES DARWIN, 1837–1858

Editor's Comments
on Papers 11, 12, and 13

11   DARWIN
Excerpt from *On Variation in a State of Nature and on the Natural Means of Selection*

12   DARWIN
Excerpt from *Natural Means of Selection*

13   DARWIN
Excerpt from *On the Variation of Organic Beings in a State of Nature; on the Natural Means of Selection; On the Comparison of Domestic Races and True Species*

The theory that adaptive evolution occurs as the result of selection operating on heritable individual variations was constructed in 1838 by Charles Darwin (1809–1882). Darwin had served as naturalist aboard the H.M.S. *Beagle* during its 1831–1836 world-encircling voyage. As he began to prepare his journal of the voyage for publication and to study his specimen collections, he soon realized that numerous facts indicated the common descent of species. This realization led Darwin to open the first of a series of *Notebooks on Transmutation of Species* (1837–1839, in de Beer, 1960–1967) in which he recorded "any facts that might bear on the question" (Darwin and Seward, 1903, vol. 1, p. 367). Darwin made the following entry in his diary for 1837:

> In July opened first note book on "Transmutation of Species"—Had been greatly struck from about Month of previous March on character of S. American fossils—& species on Galapagos Archipelago. These facts origin (especially latter) of all my views. (Darwin, 1838, in de Beer, 1959, p. 7)

The intellectual path Charles Darwin took between 1831 and 1838 that led him to construct his theory that selection operating on heritable variations produces adaptive evolution has been summarized by Bajema (1982, 1983). The claim that Charles Darwin merely transferred Malthus's ideas concerning struggle among humans to the rest of the animate world when he constructed his theory of adaptive evolution by selection has been rejected by Bowler (1976, pp. 647–648). Before he began reading Malthus's *Essay on the Principle of Population* in September 1838, Darwin's attention had been drawn to the "struggle for existence" in the animate world by Charles Lyell and other writers (Bajema, 1983, pp. 60–134).

Why did Malthus's *Essay on the Principle of Population* lead Darwin to immediately formulate his theory that adaptations were the outcome of selection? Darwin had read a book review of Comte's *Cours de Philosophie Positive* two months before reading Malthus's *Essay* and was attempting to find the biological equivalent of Comte's "numerical verification," which was "an almost indispensable criterion . . . of every hypothesis" in the most positive (scientific) of all the sciences—astronomy (Brewster, 1838, p. 299; see also Schweber, 1977, 1978). Darwin's search for quantitative statements in biology led him to read Malthus. He read Malthus's summary of Benjamin Franklin's views on "the constant tendency of all animated life to increase beyond nourishment," and of Franklin's conclusion that ". . . the [human] population, when unchecked, goes on doubling itself every twenty-five years, or increases in a geometric ratio" (Malthus, 1826, p. 6). Malthus, by stating that a population tends to increase in a geometric ratio, enabled Darwin to perceive just how strong an ecological force selection operating in nature could be. The power of selection was generated by the overproduction of offspring that greatly intensified the "struggle for existence" in nature (Limoges, 1970; Schweber, 1977, 1978).

The entries that Darwin made in his *Notebooks on Transmutation* enable historians to reconstruct most of the intellectual path that he took in becoming aware that not all adaptations are due to selective mortality and that selection operating in nature should be partitioned into natural selection and sexual selection. He read examples of behavior and morphology, the explanations for which he was to later construct his theory of sexual selection, in his grandfather's *Zoonomia* (Darwin, 1794), Sebright's 1809 pamphlet on *The Art of Improving the Breeds of Domestic Animals . . .*, and Hunter's "An Account of an Extraordinary Pheasant," published in *Animal Oeconomy* (Hunter 1837, pp. 44-49). A study of the extant entries that Charles Darwin made in his three *Notebooks on the Transmutation of Species* indicates that he possessed all the knowledge he needed to make the distinction between natural selection and both forms (male combat and female choice) of sexual selection prior to September 28, 1838, the time when Darwin first realized just how important a force selection operating in nature is (Darwin, 1837-1839, in de Beer, 1960-1967, see entries C61, 178; D76, 99, 113-114). Darwin explicitly mentioned both forms of what he was later to define as sexual selection in the following entry he made in his *M Notebook on Man, Mind and Materialism* in September 1838: "Jealousy probably originally entirely sexual; first try to attract female (or object of attachment) & then failing to drive away rival" (Gruber, 1974, p. 295).

Darwin also abstracted "An Account of an Extraordinary Pheasant" written by the English physician John Hunter (1837, pp. 44-49) during

September 1838, noting Hunter's division of "sexual marks into primary & secondary, the latter only being developed when the first become of use" (Darwin, 1838, in de Beer, 1960, Part III, D161). Hunter's essay helped Darwin pinpoint the problem that sexual dimorphism in secondary sexual characteristics poses for any comprehensive theory about the causes of evolution. In his account of an extraordinary hen pheasant that developed the feathers of the opposite sex, Hunter (1837, pp. 45-46) also drew attention to the fact that sexual dimorphism in "external form is more particularly remarkable in animals whose females are of a peaceable nature" and that

> the males of almost every class of animals are probably disposed
> to fight, being, as I have observed, stronger than the females; and
> in many of these there are parts destined solely for that purpose,
> as the spurs in the cock, and the horns in the bull . . . .

Darwin never used the term *selection* to describe how species were made in his 1837-1839 *Notebooks* before and after reading Malthus. Rather, he used the work "picking" to describe man's selective breeding of domesticated species, and the words "destruction," "preservation," and "sift out" to describe natural selection (see de Beer, 1960-1967, entries E71-72, E111, E118). Charles Darwin stated in his autobiography that since he was so anxious to avoid prejudice with respect to his new theory, he was determined for some time not to write even the briefest sketch of it (Barlow, 1958, p. 120).

Darwin (1859, p. 61) wrote in the *Origin* that he adopted the metaphor "natural selection" to describe the natural means "by which each slight variation, if useful, is preserved . . . in order to mark its relation to man's power of selection." He may have realized the value of selection as a metaphor in communicating his ideas when he read Walker's summary of selection applied to both domestic animals and human beings (Walker, 1839, pp. 263-265, 314-316) or one of William Youatt's books on domesticated animals, each of which contained passages on selective breeding where Youatt used the term *selection* (Youatt, 1831, 1834; see also Bajema, 1982; Vorzimmer, 1977). Walker's *Intermarriage* contained three pages (314-316) of quotations taken from Prichard (see Paper 10) and Lawrence (1823) on the effects of selection and selective mating in the human species. It is not known whether these passages inspired Darwin to divide selection operating in nature into sexual selection and natural selection. His personal notebooks for 1839-1841, which are being transcribed for publication in the near future, contain no mention of sexual selection (Kohn, 1980, personal communication). In 1871 Darwin recalled that it was the importance of the distinction between males having "acquired structure, not from being better fitted to survive in the struggle for existence, but from having gained an advantage over other males [with respect to reproduction], and from having transmitted this advantage to their male offspring alone" that led him "to designate this form of selection as sexual selection" (page 257 in Paper 16C).

The historian Michael Ruse (1979, p. 179) contends that it was probably Darwin's development of the domestic/natural analogy in the context of scientific justification for natural selection that led Darwin back to the context of discovery with respect to sexual selection sometime between 1839 and 1842. Ruse points out that man has selectively bred his domesticated species for two qualities: qualities that aid human livelihood—stronger horses, heavier cows, shaggier sheep, bigger vegetables; and qualities that give humans pleasure such as combative strength—a fiercer fighting cock—and beauty—a fancier pigeon. Ruse (1979, p. 179) also points out that Darwin's natural selection theory is analogous to humans' selecting traits that help domesticated organisms survive whereas Darwin's sexual selection theory is analogous to humans selecting traits that give them pleasure. Darwin (1859) tied both sexual selection for "combative strength" and "beauty" very tightly to the analogous forms of artificial selection in the *Origin*. This fact makes it seem highly probable that the domestic/natural analogy played a crucial role both in Darwin's isolating sexual selection as a separate force in nature and in his decision that the natural/sexual selection dichotomy was so important that he should draw explicity attention to it in his theory by creating the metaphor "sexual selection" (Ruse, 1979, p. 179).

Charles Darwin waited almost four years before he wrote the first summary of his theory of "the way in which new varieties become so exquisitely adapted to the external conditions of life and to other surrounding beings" (Darwin, 1845 letter, reprinted in Darwin, 1899, vol. 1, p. 395). This *Pencil Sketch of 1842* (Darwin, 1909) contains one paragraph (Paper 11) devoted to sexual selection theory. Darwin argued by analogy from man's selection, which produces varieties under domestication, to the circumstances that lead to the production of wild varieties in nature to establish the possibility of his theory (Manier, 1978, p. 7). Darwin introduced sexual selection in his 1842 *Sketch* by stating, "Besides selection by death [natural selection] . . . [there is] selection in time of fullest viguor, namely the struggle of males . . . struggle of war or charms . . ." (see Paper 11).

Thus by 1842 Darwin was already dividing the ecological causes of sexual selection into male power in combat—"struggle of war"—and male power to charm females. Darwin concluded that the "male which . . . is in fullest vigour, or best armed with arms or ornaments

101

of its species, will gain . . . some small advantage and transmit such characters to its offspring." He drew an analogy with artificial selection by comparing sexual selection "to man using male alone of good breed" and concluded his discussion of sexual selection by stating that "This latter section only of limited application, applies to variation of sexual characters" (See Paper 11).

Two years later, in 1844, Darwin wrote an expanded manuscript summarizing his theory and he made arrangements for its publication in case of his death. This *Essay of 1844* (Darwin, 1909) contains a paragraph on sexual selection (Paper 12), which is essentially an expanded version of the paragraph he had written two years earlier. Darwin did add a couple of ideas, however.

First, he pointed out that sexual selection favors adaptations that benefit the individuals possessing them but that do not benefit the species, that is, favors "the selection of individual forms, no way related to their power of obtaining food, or of defending themselves from their natural enemies, but of fighting one with another." Nonetheless, he also thought that sexual selection benefited the species because "the most vigorous males, implying perfect adaptation, must generally gain the victory in their several contests." Second, Darwin emphasized the difference between how natural selection and sexual selection produce adaptations—"This kind of selection [sexual], however, is less rigorous than the other [natural selection]; it does not require the death of the less successful, but gives to them fewer descendants" (Paper 12).

Ten years later, in 1854, Charles Darwin finally began preparing for publication his notes on his theory concerning the origin of species by selection. His receipt of Alfred Wallace's manuscript "On the Tendency of Varieties to depart indefinitely from the Type" in June 1858 triggered a chain of events that led Darwin to publish a five paragraph extract of his long manuscript on species in 1858 and the *Origin of Species,* an abstract of his long manuscript on natural selection in 1859.

Darwin considered sexual selection to be such an integral and important part of his theory about the ecological causes and adaptive consequences of selection that sexual selection was the subject of one of the five paragraphs he extracted from his long manuscript on species to be read before the Linnean Society at the same time Wallace's paper was presented by Joseph Hooker and Charles Lyell. This paragraph on sexual selection (Paper 13), which was published in 1858, is virtually identical to the paragraph (Paper 12) Darwin wrote in his *Essay of 1844.* Darwin added the adjective "secondary" to sexual characters and deleted the word "natural" from "this natural struggle amongst males."

The analysis of Charles Darwin's written thoughts on sexual selection theory reprinted in this volume indicates that Darwin had developed the crucial elements of his sexual selection theory before he wrote the first sketch of his theory in 1842!

REFERENCES

Bajema, C. J., ed., 1982, *Artificial Selection and the Development of Evolutionary Theory,* Benchmark Papers in Systematic and Evolutionary Biology, vol. 4, Hutchinson Ross, Stroudsburg, Pa., 361p.

Bajema, C. J., ed., 1983, *Natural Selection Theory: From the Speculations of the Greeks to the Quantitative Measurements of the Biometricians,* Benchmark Papers in Systematic and Evolutionary Biology, vol. 5, Hutchinson Ross, Stroudsburg, Pa., 384p.

Barlow, N., ed., 1958, *The Autobiography of Charles Darwin,* Harcourt, Brace and World, New York, 253p. (Original omissions restored, appendix and notes added.)

Bowler, P. J., 1976, Malthus, Darwin, and the Concept of Struggle, *J. Hist. Ideas* **37**:631–650.

Brewster, D. (Anonymous), 1838, Review of Comte's "Cours de Philosophie Positive," *Edinburgh Rev.* **67**:271–308.

Darwin, C. R., 1858, Extract from an Unpublished Work on Species, by C. Darwin, Esq., Consisting of a Portion of a Chapter Entitled, "On Variation of Organic Beings in a State of Nature; On the Natural Means of Selection; On the Comparison of Domestic Races and True Species, *Linn. Soc. Lond. (Zool.) J.* **3**:46–50. (See also Paper 16 in Bajema, 1983, pp. 187–191.)

Darwin, C. R., 1859, *On the Origin of Species by Means of Natural Selection or the Preservation of Favoured Races in the Struggle for Life,* 1st ed., J. Murray, London, 502p. (facsimile edition by Harvard University Press, 1966).

Darwin, E., 1794, *Zoonomia, or the Laws of Organic Life,* vol. 1, Johnson, London, 586p.

Darwin, F., ed., 1909, *The Foundations of the Origin of Species; Two Essays Written in 1842 and 1844, by Charles Darwin,* Cambridge University Press, Cambridge, England, 263p. (Both essays have been reprinted in *Evolution by Selection,* G. de Beer, ed., Cambridge University Press, Cambridge, England, 1958.)

Darwin, F., and A. C. Seward, eds., 1903, *More Letters of Charles Darwin,* 2 vols., Appleton, New York, 494p., 508p.

de Beer, G., ed., 1959, Darwin's Journal [Diary], *Br. Mus. (Nat. Hist.) Bull. Hist. Ser.* **2**(1): 1–21.

de Beer, G., ed., 1960–1967, *Darwin's Notebooks on Transmutation of Species.*
Part I. First Notebook (July 1837–February 1838) (Notebook B), *Br. Mus. (Nat. Hist.) Bull. Hist. Ser.* **2**(2):23–73 (1960).
Part II. Second Notebook (February 1838–July 1838) (Notebook C), *Br. Mus. (Nat. Hist.) Bull. Hist. Ser.* **2**(3):75–118 (1960).
Part III. Third Notebook (July 15, 1838–October 2, 1838) (Notebook D), *Br. Mus. (Nat. Hist.) Bull. Hist. Ser.* **2**(4): 119–150 (1960).

Part IV. Fourth Notebook (October 1838-July 10, 1839) (Notebook E), *Br. Mus. (Nat. Hist.) Bull. Hist. Ser.* **2**(5): 151–183 (1960).

Part V. Addenda and Corrigenda, *Br. Mus. (Nat. Hist.) Bull. Hist. Ser.* **2**(6): 185–200 (1961).

Part V. Pages Excised by Darwin, *Br. Mus. (Nat. Hist.) Bull. Hist. Ser.* **3**(5): 129–176 (1967).

Gruber, H. E., 1974, *Darwin on Man: A Psychological Study of Scientific Creativity Together with Darwin's Early and Unpublished Notebooks Transcribed and Annotated by Paul H. Barrett,* Dutton, New York, 495p.

Hunter, J., 1837, *Observations on Certain Parts of The Animal Oeconomy,* J. F. Palmer, ed., The Works of John Hunter, F.R.S. with Notes, vol. 4, Longman, Orme, Brown, Green, and Longmans, London, 506p. (and 36p. index).

Lawrence, W., 1823, *Lectures on Comparative Anatomy, Physiology, and Zoology, and the Natural History of Man,* R. Carlile, London (reprinted as Paper 8 in Bajema, 1982, pp. 60–63).

Malthus, T. R., 1826, *An Essay on the Principle of Population; or, A View of Its Past and Present Effects on Human Happiness; With an Inquiry into Our Prospects Respecting the Future Removal or Mitigation of the Evils Which It Occasions,* 6th ed., 2 vols., J. Murray, London, 535p., 528p. (See Bajema, 1983, for a facsimile reprint of Malthus's summary of Benjamin Franklin's views.)

Manier, E., 1978, *The Young Darwin and His Cultural Circle: A Study of Influences Which Helped Shape the Language and Logic of the First Drafts of the Theory of Natural Selection,* D. Reidel, Boston, 242p.

Ruse, M., 1979, *The Darwinian Revolution: Science Red in Tooth and Claw,* University of Chicago Press, Chicago, 320p.

Schweber, S. S., 1977, The Origin of the *Origin* Revisited, *J. Hist. Biol.* **10:**220–316.

Schweber, S. S., 1978, The Genesis of Natural Selection—1838: Some Further Insights, *BioScience* **28:**321–326.

Vorzimmer, P., 1975, An Early Darwin Manuscript: The "Outline and Draft of 1839, " *J. Hist. Biol.* **8:**191–217.

Walker, A., 1839, *Intermarriage: or The Mode in Which, and the Causes Why, Beauty, Health and Intellect, Result from Certain Unions, and Deformity, Disease and Insanity from Others: . . . and . . . an Account of Corresponding Effects in the Breeding of Animals,* J. & H. Langley, New York, 384p.

Wallace, A. R., 1858, On the Tendency of Varieties to Depart Indefinitely from the Original Type, *Linn. Soc. Lond. (Zool.) J.* **3:**53–62 (reprinted as part of Paper 16 in Bajema, 1983).

Youatt, W., 1831, *The Horse: With a Treatise on Draught; and a Copius Index,* Baldwin and Cradock, London, 472p.

Youatt, W., 1834, *Cattle: Their Breeds, Management, and Diseases,* Baldwin and Cradock, London, 600p.

11

ON VARIATION IN A STATE OF NATURE AND ON THE NATURAL MEANS OF SELECTION (1842)

C. R. Darwin

[*Editor's Note:* In the original, material precedes and follows this excerpt.]

Besides selection by death, in bisexual animals (illegible) the selection in time of fullest vigour, namely struggle of males; even in animals which pair there seems a surplus (?) and a battle, possibly as in man more males produced than females, struggle of war or charms[1]. Hence that male which at that time is in fullest vigour, or best armed with arms or ornaments of its species, will gain in hundreds of generations some small advantage and transmit such characters to its offspring. So in female rearing its young, the most vigorous and skilful and industrious, (whose) instincts (are) best developed, will rear more young, probably possessing her good qualities, and a greater number will thus (be) prepared for the struggle of nature. Compared to man using a male alone of good breed. This latter section only of limited application, applies to variation of [specific] sexual characters. Introduce here contrast with Lamarck,—absurdity of habit, or chance ?? or external conditions, making a woodpecker adapted to tree[2].

---

[1] Here we have the two types of sexual selection discussed in the *Origin*, Ed. i. pp. 88 et seq., vi. pp. 108 et seq.

[2] It is not obvious why the author objects to "chance" or "external conditions making a woodpecker." He allows that variation is ultimately referable to conditions and that the nature of the connexion is unknown, i.e. that the result is fortuitous. It is not clear in the original to how much of the passage the two ? refer.

12

NATURAL MEANS OF SELECTION (1844)

C. R. Darwin

[*Editor's Note:* In the original, material precedes and follows this excerpt.]

Besides this natural means of selection, by which those individuals are preserved, whether in their egg or seed or in their mature state, which are best adapted to the place they fill in nature, there is a second agency at work in most bisexual animals tending to produce the same effect, namely the struggle of the males for the females. These struggles are generally decided by the law of battle;

but in the case of birds, apparently, by the charms of their song[1], by their beauty or their power of courtship, as in the dancing rock-thrush of Guiana. Even in the animals which pair there seems to be an excess of males which would aid in causing a struggle: in the polygamous animals[2], however, as in deer, oxen, poultry, we might expect there would be severest struggle: is it not in the polygamous animals that the males are best formed for mutual war? The most vigorous males, implying perfect adaptation, must generally gain the victory in their several contests. This kind of selection, however, is less rigorous than the other; it does not require the death of the less successful, but gives to them fewer descendants. This struggle falls, moreover, at a time of year when food is generally abundant, and perhaps the effect chiefly produced would be the alteration of sexual characters, and the selection of individual forms, no way related to their power of obtaining food, or of defending themselves from their natural enemies, but of fighting one with another. This natural struggle amongst the males may be compared in effect, but in a less degree, to that produced by those agriculturalists who pay less attention to the careful selection of all the young animals which they breed and more to the occasional use of a choice male[3].

[1] These two forms of sexual selection are given in *Origin*, Ed. i. p. 87, vi. p. 107. The Guiana rock-thrush is given as an example of bloodless competition.

[2] (Note in original.) Seals? Pennant about battles of seals.

[3] In the Linnean paper of July 1, 1858 the final word is *mate*: but the context shows that it should be *male*; it is moreover clearly so written in the MS.

13

Reprinted from *Linn. Soc. London (Zool.) J.* **3**:50 (1858)

ON THE VARIATION OF ORGANIC BEINGS IN A STATE OF NATURE; ON THE NATURAL MEANS OF SELECTION; ON THE COMPARISON OF DOMESTIC RACES AND TRUE SPECIES

C. R. Darwin

[*Editor's Note:* In the original, material precedes this excerpt.]

Besides this natural means of selection, by which those individuals are preserved, whether in their egg, or larval, or mature state, which are best adapted to the place they fill in nature, there is a second agency at work in most unisexual animals, tending to produce the same effect, namely, the struggle of the males for the females. These struggles are generally decided by the law of battle, but in the case of birds, apparently, by the charms of their song, by their beauty or their power of courtship, as in the dancing rock-thrush of Guiana. The most vigorous and healthy males, implying perfect adaptation, must generally gain the victory in their contests. This kind of selection, however, is less rigorous than the other; it does not require the death of the less successful, but gives to them fewer descendants. The struggle falls, moreover, at a time of year when food is generally abundant, and perhaps the effect chiefly produced would be the modification of the secondary sexual characters, which are not related to the power of obtaining food, or to defence from enemies, but to fighting with or rivalling other males. The result of this struggle amongst the males may be compared in some respects to that produced by those agriculturists who pay less attention to the careful selection of all their young animals, and more to the occasional use of a choice mate.

Part III

SEXUAL SELECTION THEORY IN THE ORIGIN OF SPECIES, 1859–1872

Editor's Comments
on Papers 14 and 15

In all six editions of his *On the Origin of Species by Means of Natural Selection, or the Preservation of Favoured Races in the Struggle for Life* (1859–1872),* Charles Darwin contended that three types of selection were involved in bringing about "descent with modification." He divided the ecological interactions an organism has that cause evolution into the following three categories: "man's power of selection" (artificial selection, see Bajema, 1982), natural selection (see Bajema, 1983) and sexual selection (Fig. 1). By dividing selection into these three categories, Darwin was able, first, to argue by analogy from artificial selection—a process already widely known to bring about hereditary changes in domesticated species—to both natural and sexual selection, which were operating in the rest of nature (Figs. 2 and 3). Second, the division of selection into these three categories enabled Darwin to partition those intraspecific interactions that selected for adaptations, which were advantageous to the individuals possessing them but harmful to certain other members of the species into a separate selective system—sexual selection. Darwin's discussion of sexual selection in the first edition of the *Origin* is reprinted in Papers 14A, 14B, 14C, and 14D. He argued by analogy from artificial selection to natural selection and to both forms of sexual selection—

*All the changes in text that Charles Darwin made in the second and later editions have been catalogued by Peckham (1959).

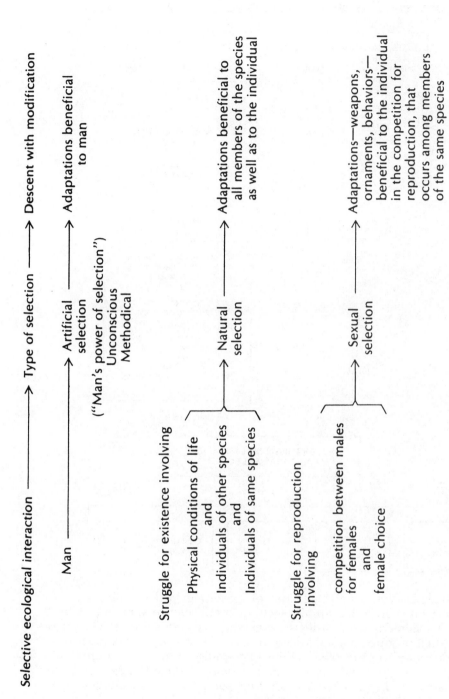

Figure 1. Charles Darwin's 1859 system for classifying selection. (After G. Hardin and C. Bajema, *Biology: Its Principles and Implications*, 3rd ed. San Francisco: W. H. Freeman, 1978.)

111

THE LOGIC OF CHARLES DARWIN'S THEORY OF NATURAL SELECTION

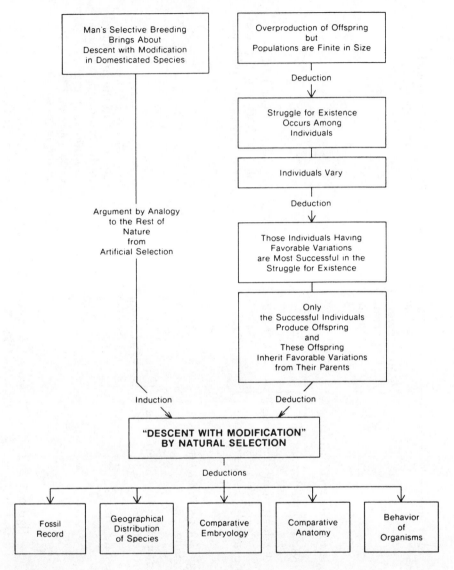

Figure 2. Darwin used three lines of reasoning in developing and advocating his theory of natural selection: the inductive vera causa ("true cause") argument from artificial selection; the deduction of the set of causes that produce natural selection; and the rationalist vera causa used to deduce the kinds of observations that would be made in nature if natural selection operated the way that Darwin predicted. (After M. Ruse, *The Darwinian Revolution: Science Red Tooth and Claw,* Chicago: University of Chicago Press, 1979, p. 198.)

112

THE LOGIC OF CHARLES DARWIN'S THEORY OF SEXUAL SELECTION

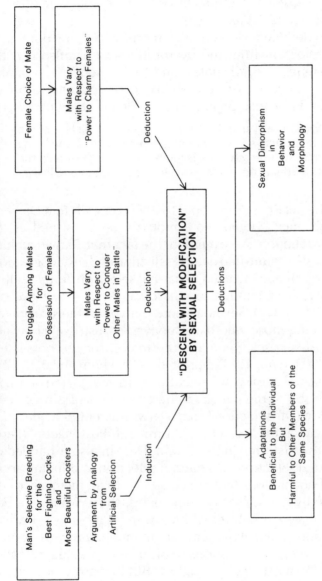

Figure 3. Darwin also used three lines of reasoning in developing and advocating his theory of sexual selection: the inductive vera causa ("true cause") argument from artificial selection; the deduction of the set of natural causes that produce sexual selection; and the rationalist vera causa used to deduce the kinds of observations that would be made in nature if sexual selection operated the way that Darwin predicted. (After M. Ruse, *The Darwinian Revolution: Science Red Tooth and Claw*, Chicago: University of Chicago Press, 1979, p. 198.)

113

male combat and female choice (Paper 14A, pp. 88–89; see also Figs. 2 and 3 and Bajema, 1882). Darwin emphasized that: "I have called this principle, by which each slight variation, if useful, is preserved, by the term of Natural Selection, in order to mark its relation to man's power of selection" (1859, p. 61).

Darwin also deduced "descent with modification" by both natural and sexual selection from the overproduction of offspring, the struggle for existence and for mates, and the presence of heritable variations with respect to both the ability to survive and to obtain mates (Fig. 2 and 3, see also Bajema, 1983). He pointed out:

> I should premise that I use the term Struggle for Existence in a large and metaphorical sense, including dependence of one being on another, and including (which is more important) not only the life of the individual, but success in leaving progeny. (Darwin, 1859, p. 62)

The historian Michael Ruse (1971, pp. 316-317) contends that Darwin's "struggle for existence" would more properly be construed as the "struggle for reproduction" were it not for the fact that Darwin divided selection operating in nature into natural selection and sexual selection. The struggle for existence involved in natural selection "is basically a struggle to survive long enough to *reproduce,* that is, to leave offspring (Ruse, 1971, p. 317). Sexual selection involves the struggle for reproduction among those who have survived—"a struggle between the individuals of one sex, generally the males, for the possession of the other sex" (Darwin, 1872, reprinted in Peckham, 1959, p. 173).

Darwin's discussion of sexual selection in the *Origin* is essentially an expanded version of the paragraph he wrote in his *Essay of 1844.* The metaphor "sexual selection" that Darwin invented to describe what he had been calling "selection in time of fullest vigour" (Paper 11) appears for the first time in the *Origin*. Darwin also expanded on his idea that sexual selection aids natural selection in bringing about the evolution of adaptations that are beneficial to all members of the species: "Generally, the most vigorous males, those which are best fitted for their places in nature, will leave most progeny" (Darwin, 1859, p. 88 in Paper 14A). However, Darwin immediately went on to state that there are many exceptions to his idea that sexual selection generally favors the most vigorous males: "But in many cases, victory will depend not on general vigour, but on having special weapons, confined to the male sex" (p. 88 in Paper 14A). He modified this sentence in the last (1872) edition of the *Origin* to read: "Victory depends not so much on general vigour, as on having special weapons . . ." (reprinted in Peckham, 1959, p. 174).

Darwin discussed the role of female choice of male charms

among birds in all six editions of the *Origin:* "Successive males display their gorgeous plumage and perform strange antics before the females, which standing by as spectators, at last choose the most attractive partner" (Darwin, 1859–1872, reprinted in Peckham, 1959, p. 175). Likewise in all six editions he contended that sexual selection was chiefly responsible for the differences between the races of man (Darwin, 1859, p. 199, in Paper 14C).

In the fourth edition of the *Origin,* Darwin added an attack on "the view that organic beings have been created beautiful for the delight of man . . ." in his chapter that dealt with difficulties of his theory. Darwin admitted that some beauty had arisen merely for beauty's sake. However, he contended that such beauty had been brought into existence "not for the delight of man, but through sexual selection, this is from the more beautiful males having been continually preferred by their less ornamented females" (Darwin, 1866, reprinted in Peckham, 1959, pp. 369–372).

Darwin did revise the sixth edition of the *Origin* to take into account the existence of sex-reversed species, in which the females compete with each other for males and males choose females, by changing the phrase "a struggle between the males for possession of the females" to read "a struggle between the individuals of one sex, generally the males, for the possession of the other sex" (Darwin, 1872, reprinted in Peckham, 1959, pp. 173–174).

He also devoted three pages of his "Laws of Variation" chapter in all editions of the *Origin* to a discussion of the variability of secondary sexual characters and the accumulation of such variations by both natural and sexual selection for secondary and ordinary specific purposes (pp. 156–158 in Paper 14B).

The existence of sterile castes of social insects constitutes a problem that any comprehensive theory of evolution must solve. At first, Darwin thought that the existence of neuter or sterile females was fatal to his theory of adaptive evolution by selection because they cannot propagate their kind (p. 236 in Paper 14D). Darwin may well have procrastinated between 1844 and 1856 with respect to writing the *Origin* because he had not yet solved this problem (Richards, 1983).

Darwin employed an argument by analogy from artificial selection to solve the neuter insect problem. He contended that "selection may be applied to the family, as well as to the individual, and may thus gain the desired end" (p. 237 in Paper 14D). He went on to point out that neuter insects are a fatal difficulty for Lamarckian theories of evolution since "no amount of exercise, or habit, or volition, in the utterly sterile members of a community could possibly have affected the structure or instincts of the fertile members which alone leave descendants" (Darwin, 1859, p. 242).

A century later Darwin's solution was generalized and quantified by W. D. Hamilton (1964) who constructed the inclusive fitness theory of selection, which is the theoretical foundation for much of sociobiology. Kin selection theory, as inclusive fitness theory is widely known, expands our scientific understanding of selection theory by taking into account the ways in which an individual furthers its own selfish genetic interest (maximizing the reproduction of copies of the genes it is carrying) via the reproduction of its genetic kin who are carrying genes identical by descent to those carried by the individual organism.

Darwin viewed natural selection as bringing about evolution primarily by affecting the survival of individuals before or during reproductive age. Sexual selection brought about evolution primarily by affecting the reproduction of those individuals who survived to reproductive age. When one takes Darwin's theory of sexual selection seriously, it is easier to understand why in 1866 Darwin adopted Herbert Spencer's immortal metaphor "Survival of the Fittest" to describe natural selection (Bajema, 1983, Part VI). Darwin never used the phrase "survival of the fittest" to describe sexual selection. The scientific community's opposition to Darwin's theory of sexual selection and the ambiguity of Spencer's metaphor have both been responsible to a great extent for widespread misconceptions about the role that selection plays in bringing about and maintaining adaptations (Bajema, 1983, Part VI).

After the publication of the *Origin,* Charles Darwin vigorously pursued his sexual selection theory in his personal correspondence by raising the question with several naturalists—especially Alfred Wallace and Henry Bates. Why was Darwin so interested in sexual selection? The biologist and historian Michael Ghiselin (1974, p. 176) has contended:

> Evolution by orthogenesis, or even by strictly natural selection, left open the possibility of divine providence and foresight, and was therefore not entirely unpalatable. But sexual selection allowed for nothing of the sort. Nature, in producing contraceptions rather than contrivances, and acting against the interests of the species, gives rise to a spectacle of purposelessness and triviality. Darwin realized this, and much of his work on sex was explicitly directed against teleology.

The naturalist Henry Bates wrote the following letter to Charles Darwin on March 28, 1861, in response to a query about sexual selection:

> In the Aenaes section of the Genus Papilio, the males are generally of extremely brilliant colours . . . the females are plainer & so different from the males that they were generally held to be

distinct species until I took them in copula . . . I thoroughly believe in your theory of sexual selection. (Bates, 1861, reprinted in Stecher, 1969, p. 13)

Darwin wrote Bates on April 4:

Hardly anything in your letter has pleased me more than about sexual selection. . . . If I had to cut up myself in a Review, I would have worked & quizzed Sexual Selection: therefore, though I am fully convinced that it is largely true, you may imagine how pleased I am at what you say are your beliefs.—This part of your letter to me is a quintessence of richness. (Darwin and Seward, 1903, vol. 1, pp. 182–183)

The debate between Darwin and Wallace over the relationship between sexual selection and natural selection in the origin of sexual dimorphism and, in particular, the differences in coloration between the sexes has been reviewed by Kottler (1980). In his correspondence with Wallace, Darwin championed his theory that female choice was the cause of sexual dimorphism in those species of butterflies and birds where the male is more brightly colored than the female.

Darwin also asked Wallace numerous questions about the occurrence of sexual dimorphism. These queries led Wallace to study more carefully those species in which the sexes differ considerably, especially with respect to color. Consequently Wallace (1) reported the first case of sex-limited mimicry in the literature when he reported that there were species among the swallowtail butterflies in which the female alone was brightly colored, mimicking another species of butterfly that was unpalatable (Wallace, 1865), and (2) introduced the problem of sex reversed species—species in which the male spends more time rearing the offspring than the female—into the debate over the importance of female choice (Paper 15). Darwin revised the sixth edition of the *Origin* to allow for female-female competition and male choice of female mates in sex-reversed species in which the male makes the larger parental investment:

This form of selection (sexual) depends, not on a struggle for existence in relation to other organic beings or the external conditions, but on a *struggle between individuals of one sex, generally the males, for the possession of the other sex.* (Darwin, 1872, reprinted in Peckham, 1959, pp. 173–174, italics added)

Darwin renewed his efforts to apply sexual selection theory to human evolution by March 7, 1860—the date he wrote Wallace stating "I have a good many facts which make me believe in sexual selection as applied to man, but whether I shall convince anyone else is very doubtful" (Darwin, 1860, reprinted in Marchant, 1916, p. 116). When Wallace published his essay on *The Origin of Human Races and the Antiquity of Man Deduced from Natural Selection* in 1864, Darwin wrote Wallace:

I suspect that a sort of sexual selection has been the most powerful means of changing the races of man. I can show that the different races have a widely different standard of beauty. Among savages the most powerful men will have the pick of the women, and they will generally leave the most descendants. (May 28, 1864, letter reprinted in Marchant, 1916, p. 128)

Upon completing his two volumes *Variation of Animals and Plants under Domestication* (1868), Darwin made the following entry in his diary (de Beer, 1959, p. 18) for February 4, 1868: "Began on Man & Sexual Selection."

REFERENCES

Bajema, C. J., ed., 1982, *Artificial Selection and the Development of Evolutionary Theory,* Benchmark Papers in Systematic and Evolutionary Biology, vol. 4, Hutchinson Ross, Stroudsburg, Pa., 361p.

Bajema, C. J., ed., 1983, *Natural Selection Theory: From the Speculations of the Greeks to the Quantitative Measurements of the Biometricians,* Benchmark Papers in Systematic and Evolutionary Biology, vol. 5, Hutchinson Ross, Stroudsburg, Pa., 384p.

Darwin, C. R., 1859-1872, *On the Origin of Species by Means of Natural Selection or the Preservation of Favoured Races in the Struggle for Life,* 6 editions; 1st ed., J. Murray, London, 502p. (facsimile edition by Harvard University Press, 1966).

Darwin, F., and A. C. Seward, eds., 1903, *More Letters of Charles Darwin,* 2 vols., Appleton, New York, 494p., 508p.

de Beer, G., ed., 1959, "Darwin's Journal," *Br. Mus (Nat. Hist.) Bull.Hist. Ser.* **2**(1):1-21.

Ghiselin, M. T., 1974, *The Economy of Nature and the Evolution of Sex,* University of California Press, Berkeley, 346p.

Hamilton, W. D., 1964, The Genetical Evolution of Social Behaviour, Parts I and II, *J. Theor. Biol.* **7**:1-16, 17-52.

Kottler, M., 1980, Darwin, Wallace, and the Origin of Sexual Dimorphism, *Am. Philos. Soc. Proc.* **124**:203-226.

Marchant, J., ed., 1916, *Alfred Russel Wallace: Letters and Reminiscences,* Harper, New York, 507p.

Peckham, M., ed., 1959, *The Origin of Species by Charles Darwin, A Variorum Text,* University of Pennsylvania Press, Philadelphia, 816p.

Richards, R. J., 1983, Why Darwin Delayed, or Interesting Problems and Models in the History of Science, *J. Hist. Behav. Sci.* **19**:45-53.

Ruse, M., 1971, Natural Selection in "The Origin of Species," *Hist. Philos. Sci.* **1**:311-351.

Stecher, R. M., 1969, The Darwin-Bates Letters: Correspondence between Two Nineteenth-Century Travellers and Naturalists, Part I, *Ann. Sci.* **25**:1-47.

Wallace, A. R., 1864, The Origin of Human Races and the Antiquity of Man Deduced from Natural Selection, *Anthropol. Rev.* **2**:clviii-clxx.

Wallace, A. R., 1865, On the Phenomena of Variation and Geographical Distribution, as Illustrated by the Papilionidae of the Malayan Region, *Linn Soc. London Trans.* **25**:1-71.

14A

Reprinted from pages 87–90 of *On the Origin of Species by Means of Natural Selection or the Preservation of Favoured Races in the Struggle for Life,* J. Murray, London, 1859

SEXUAL SELECTION

C. R. Darwin

Sexual Selection.—Inasmuch as peculiarities often appear under domestication in one sex and become hereditarily attached to that sex, the same fact probably occurs under nature, and if so, natural selection will be able to modify one sex in its functional relations to the other sex, or in relation to wholly different habits of life in the two sexes, as is sometimes the case with insects.   And this leads me to say a few words on what I call Sexual Selection.   This depends, not on a struggle for existence, but on a struggle between the males for possession of the females ; the result is not death to the unsuccessful competitor, but few or no offspring.   Sexual selection is, therefore, less rigorous than natural selection.   Generally, the most vigorous males, those which are best fitted for their places in nature, will leave most progeny.   But in many cases, victory will depend not on general vigour, but on having special weapons, confined to the male sex.   A hornless stag or spurless cock would have a poor chance of leaving offspring.   Sexual selection by always allowing the victor to breed might surely give indomitable courage, length to the spur, and strength to the wing to strike in the spurred leg, as well as the brutal cock-fighter, who knows well that he can improve his breed by careful selection of the best cocks.   How low in the scale of nature this law of battle descends, I know not ; male alligators have been described as fighting, bellowing, and whirling round, like Indians in a war-dance, for the possession of the females ; male salmons have been seen fighting all day long ; male stag-beetles often bear wounds from the huge mandibles of other males.   The war is, perhaps, severest between the males of polygamous animals, and these seem oftenest provided with special weapons.   The males of carnivorous animals are already well armed ; though to them and to others, special means of defence may be given through means of sexual selection, as the mane to the lion, the shoulder-pad to the boar, and the hooked jaw to the male salmon ; for the shield may be as important for victory, as the sword or spear.

Amongst birds, the contest is often of a more peaceful character.   All those who have attended to the subject,

120

believe that there is the severest rivalry between the males of many species to attract by singing the females. The rock-thrush of Guiana, birds of Paradise, and some others, congregate; and successive males display their gorgeous plumage and perform strange antics before the females, which standing by as spectators, at last choose the most attractive partner. Those who have closely attended to birds in confinement well know that they often take individual preferences and dislikes: thus Sir R. Heron has described how one pied peacock was eminently attractive to all his hen birds. It may appear childish to attribute any effect to such apparently weak means: I cannot here enter on the details necessary to support this view; but if man can in a short time give elegant carriage and beauty to his bantams, according to his standard of beauty, I can see no good reason to doubt that female birds, by selecting, during thousands of generations, the most melodious or beautiful males, according to their standard of beauty, might produce a marked effect. I strongly suspect that some well-known laws with respect to the plumage of male and female birds, in comparison with the plumage of the young, can be explained on the view of plumage having been chiefly modified by sexual selection, acting when the birds have come to the breeding age or during the breeding season; the modifications thus produced being inherited at corresponding ages or seasons, either by the males alone, or by the males and females; but I have not space here to enter on this subject.

Thus it is, as I believe, that when the males and females of any animal have the same general habits of life, but differ in structure, colour, or ornament, such differences have been mainly caused by sexual selection; that is, individual males have had, in successive generations, some slight advantage over other

121

males, in their weapons, means of defence, or charms; and have transmitted these advantages to their male offspring. Yet, I would not wish to attribute all such sexual differences to this agency: for we see peculiarities arising and becoming attached to the male sex in our domestic animals (as the wattle in male carriers, horn-like protuberances in the cocks of certain fowls, &c.), which we cannot believe to be either useful to the males in battle, or attractive to the females. We see analogous cases under nature, for instance, the tuft of hair on the breast of the turkey-cock, which can hardly be either useful or ornamental to this bird;—indeed, had the tuft appeared under domestication, it would have been called a monstrosity.

14B

Reprinted from pages 156–158 of *On the Origin of Species by Means of Natural Selection or the Preservation of Favoured Races in the Struggle for Life,* J. Murray, London, 1859

LAWS OF VARIATION

C. R. Darwin

[*Editor's Note:* In the original, material precedes this excerpt.]

I think it will be admitted, without my entering on details, that secondary sexual characters are very variable; I think it also will be admitted that species of the same group differ from each other more widely in their secondary sexual characters, than in other parts of their organisation; compare, for instance, the amount of difference between the males of gallinaceous birds, in which secondary sexual characters are strongly displayed, with the amount of difference between their females; and the truth of this proposition will be granted. The cause of the original variability of secondary sexual characters is not manifest; but we can see why these characters should not have been rendered as constant and uniform as other parts of the organisation; for secondary sexual characters have been accumulated by sexual selection, which is less rigid in its action than ordinary selection, as it does not entail death, but only gives fewer offspring to the less favoured males.   Whatever the cause may be of the variability of secondary sexual characters, as they are highly variable, sexual selection will have had a wide scope for action, and may thus readily have succeeded in giving to the species of the same group a greater amount of difference in their sexual characters, than in other parts of their structure.

It is a remarkable fact, that the secondary sexual differences between the two sexes of the same species are generally displayed in the very same parts of the organisation in which the different species of the same genus differ from each other.   Of this fact I will give in illustration two instances, the first which happen to stand on my list; and as the differences in these cases are of a very unusual nature, the relation can hardly be accidental.   The same number of joints in the tarsi is a character generally common to very large groups of beetles, but in the Engidæ, as Westwood has remarked, the number varies greatly; and the number likewise differs in the two sexes of the same species: again in fossorial hymenoptera, the manner of neuration of the wings is a character of the highest importance, because common to large groups; but in certain genera the neuration differs in the different species, and likewise in the two sexes of the same species. This relation has a clear meaning on my view of the subject: I look at all the species of the same genus as having as certainly descended from the same progenitor, as have the two sexes of any one of the species.   Consequently, whatever part of the structure of the common progenitor, or of its early descendants, became variable; variations of this part would, it is highly probable, be taken advantage of by natural and sexual selection, in order to fit the several species to their several places in the economy of nature, and likewise to fit the two sexes of the same species to each other, or to fit the males and females to different habits of life, or the males to struggle with other males for the possession of the females.

Finally, then, I conclude that the greater variability of specific characters, or those which distinguish species from species, than of generic characters, or those which the species possess in common ;—that the frequent extreme variability of any part which is developed in a species in an extraordinary manner in comparison with the same part in its congeners ; and the not great degree of variability in a part, however extraordinarily it may be developed, if it be common to a whole group of species ;—that the great variability of secondary sexual characters, and the great amount of difference in these same characters between closely allied species ;—that secondary sexual and ordinary specific differences are generally displayed in the same parts of the organisation,—are all principles closely connected together. All being mainly due to the species of the same group having descended from a common progenitor, from whom they have inherited much in common,—to parts which have recently and largely varied being more likely still to go on varying than parts which have long been inherited and have not varied,—to natural selection having more or less completely, according to the lapse of time, overmastered the tendency to reversion and to further variability,—to sexual selection being less rigid than ordinary selection,—and to variations in the same parts having been accumulated by natural and sexual selection, and thus adapted for secondary sexual, and for ordinary specific purposes.

14c

Reprinted from pages 194–199 of *On the Origin of Species by Means of Natural Selection or the Preservation of Favoured Races in the Struggle for Life*, J. Murray, London, 1859

ORGANS OF LITTLE APPARENT IMPORTANCE

C. R. Darwin

Organs of little apparent importance.—As natural selection acts by life and death,—by the preservation of individuals with any favourable variation, and by the destruction of those with any unfavourable deviation of structure,—I have sometimes felt much difficulty in understanding the origin of simple parts, of which the importance does not seem sufficient to cause the preservation of successively varying individuals. I have sometimes felt as much difficulty, though of a very different kind, on this head, as in the case of an organ as perfect and complex as the eye.

In the first place, we are much too ignorant in regard to the whole economy of any one organic being, to say what slight modifications would be of importance or not. In a former chapter I have given instances of most trifling characters, such as the down on fruit and the colour of the flesh, which, from determining the attacks of insects or from being correlated with constitutional differences, might assuredly be acted on by natural selection. The tail of the giraffe looks like an artificially constructed fly-flapper; and it seems at first incredible that this could have been adapted for its present purpose by successive slight modifications, each better and better, for so trifling an object as driving away flies; yet we should pause before being too positive even in this case, for we know that the distribution and existence of cattle and other animals in South America absolutely depends on their power of resisting the attacks of insects: so that individuals which could by any means defend themselves from these small enemies, would be able to range into new pastures and thus gain a great advantage. It is not that the larger quadrupeds are actually destroyed (except in some rare cases) by the flies, but they are incessantly harassed and their strength reduced, so that they are more subject to disease, or not so well enabled in a coming dearth to search for food, or to escape from beasts of prey.

Organs now of trifling importance have probably in some cases been of high importance to an early progenitor, and, after having been slowly perfected at a former period, have been transmitted in nearly the same state, although now become of very slight use; and any actually injurious deviations in their structure will always have been checked by natural selection. Seeing how important an organ of locomotion the tail is in most aquatic animals, its general presence and use for many purposes in so many land animals, which in their lungs or modified swimbladders betray their aquatic origin, may perhaps be thus accounted for. A well-developed tail having been formed in an aquatic animal, it might subsequently come to be worked in for all sorts of purposes, as a fly-flapper, an organ of prehension, or as an aid in turning, as with the dog, though the aid must be slight, for the hare, with hardly any tail, can double quickly enough.

In the second place, we may sometimes attribute importance to characters which are really of very little importance, and which have originated from quite secondary causes, independently of natural selection. We should remember that climate, food, &c., probably have some little direct influence on the organisation; that characters reappear from the law of reversion; that correlation of growth will have had a most important influence in modifying various structures; and finally, that sexual selection will often have largely modified the external characters of animals having a will, to give one male an advantage in fighting with another or in charming the females. Moreover when a modification of structure has primarily arisen from the above or other unknown causes, it may at first have been of no advantage to the species, but may subsequently have been taken advantage of by the descendants of the species under new conditions of life and with newly acquired habits.

To give a few instances to illustrate these latter

remarks. If green woodpeckers alone had existed, and we did not know that there were many black and pied kinds, I dare say that we should have thought that the green colour was a beautiful adaptation to hide this tree-frequenting bird from its enemies; and consequently that it was a character of importance and might have been acquired through natural selection; as it is, I have no doubt that the colour is due to some quite distinct cause, probably to sexual selection. A trailing bamboo in the Malay Archipelago climbs the loftiest trees by the aid of exquisitely constructed hooks clustered around the ends of the branches, and this contrivance, no doubt, is of the highest service to the plant; but as we see nearly similar hooks on many trees which are not climbers, the hooks on the bamboo may have arisen from unknown laws of growth, and have been subsequently taken advantage of by the plant undergoing further modification and becoming a climber. The naked skin on the head of a vulture is generally looked at as a direct adaptation for wallowing in putridity; and so it may be, or it may possibly be due to the direct action of putrid matter; but we should be very cautious in drawing any such inference, when we see that the skin on the head of the clean-feeding male turkey is likewise naked. The sutures in the skulls of young mammals have been advanced as a beautiful adaptation for aiding parturition, and no doubt they facilitate, or may be indispensable for this act; but as sutures occur in the skulls of young birds and reptiles, which have only to escape from a broken egg, we may infer that this structure has arisen from the laws of growth, and has been taken advantage of in the parturition of the higher animals.

We are profoundly ignorant of the causes producing slight and unimportant variations; and we are immediately made conscious of this by reflecting on the differences in the breeds of our domesticated animals in different countries,—more especially in the less civilized countries where there has been but little artificial selection. Careful observers are convinced that a damp climate affects the growth of the hair, and that with the hair the horns are correlated. Mountain breeds always differ from lowland breeds; and a mountainous country would probably affect the hind limbs from exercising them more, and possibly even the form of the pelvis; and then by the law of homologous variation, the front limbs and even the head would probably be affected. The shape, also, of the pelvis might affect by pressure the shape of the head of the young in the womb. The laborious breathing necessary in high regions would, we have some reason to believe, increase the size of the chest; and again correlation would come into play. Animals kept by savages in different countries often have to struggle for their own subsistence, and would be exposed to a certain extent to natural selection, and individuals with slightly different constitutions would succeed best under different climates; and there is reason to believe that constitution and colour are correlated. A good observer, also, states that in cattle susceptibility to the attacks of flies is correlated with colour, as is the liability to be poisoned by certain plants; so that colour would be thus subjected to the action of natural selection. But we are far too ignorant to speculate on the relative importance of the several known and unknown laws of variation; and I have here alluded to them only to show that, if we are unable to account for the characteristic differences of our domestic breeds, which nevertheless we generally admit to have arisen through ordinary generation, we ought not to lay too much stress on our ignorance of the precise cause of the slight analogous differences between species.   I might have adduced for this same purpose the differences between the races of man, which are so strongly marked; I may add that some little light can apparently be thrown on the origin of these differences, chiefly through sexual selection of a particular kind, but without here entering on copious details my reasoning would appear frivolous.

The foregoing remarks lead me to say a few words on the protest lately made by some naturalists, against the utilitarian doctrine that every detail of structure has been produced for the good of its possessor.   They believe that very many structures have been created for beauty in the eyes of man, or for mere variety.   This doctrine, if true, would be absolutely fatal to my theory.   Yet I fully admit that many structures are of no direct use to their possessors.   Physical conditions probably have had some little effect on structure, quite independently of any good thus gained.   Correlation of growth has no doubt played a most important part, and a useful modification of one part will often have entailed on other parts diversified changes of no direct use.   So again characters which formerly were useful, or which formerly had arisen from correlation of growth, or from other unknown cause, may reappear from the law of reversion, though now of no direct use.   The effects of sexual selection, when displayed in beauty to charm the females, can be called useful only in rather a forced sense.   But by far the most important consideration is that the chief part of the organisation of every being is simply due to inheritance; and consequently, though each being assuredly is well fitted for its place in nature, many structures now have no direct relation to the habits of life of each species.

[*Editor's Note:* Material has been omitted at this point.]

14D

Reprinted from pages 235–239 of *On the Origin of Species by Means of Natural Selection or the Preservation of Favoured Races in the Struggle for Life*, J. Murray, London, 1859

NEUTER INSECTS

C. R. Darwin

[*Editor's Note:* In the original, material precedes and follows this excerpt.]

No doubt many instincts of very difficult explanation could be opposed to the theory of natural selection, —cases, in which we cannot see how an instinct could possibly have originated; cases, in which no intermediate gradations are known to exist; cases of instinct of apparently such trifling importance, that they could hardly have been acted on by natural selection; cases of instincts almost identically the same in animals so remote in the scale of nature, that we cannot account for their similarity by inheritance from a common parent, and must therefore believe that they have been acquired by independent acts of natural selection. I will not here enter on these several cases, but will confine myself to one special difficulty, which at first appeared to me insuperable, and actually fatal to my whole theory. I allude to the neuters or sterile females in insect-communities: for these neuters often differ widely in instinct and in structure from both the males and fertile females, and yet, from being sterile, they cannot propagate their kind.

The subject well deserves to be discussed at great length, but I will here take only a single case, that of working or sterile ants. How the workers have been rendered sterile is a difficulty; but not much greater than that of any other striking modification of structure; for it can be shown that some insects and other articulate animals in a state of nature occasionally become sterile; and if such insects had been social, and it had been profitable to the community that a number should have been annually born capable of work, but incapable of procreation, I can see no very great difficulty in this being effected by natural selection. But I must pass over this preliminary difficulty. The great difficulty lies in the working ants differing widely from both the males and the fertile females in structure, as in the shape of the thorax and in being destitute of wings and sometimes of eyes, and in instinct. As far as instinct alone is concerned, the prodigious difference in this respect between the workers and the perfect females, would have been far better exemplified by the hive-bee. If a working ant or other neuter insect had been an animal in the ordinary state, I should have unhesitatingly assumed that all its characters had been slowly acquired through natural selection; namely, by an individual having been born with some slight profitable modification of structure, this being inherited by its offspring, which again varied and were again selected, and so onwards. But with the working ant we have an insect differing greatly from its parents, yet absolutely sterile; so that it could never have transmitted successively acquired modifications of structure or instinct to its progeny. It may well be asked how is it possible to reconcile this case with the theory of natural selection?

First, let it be remembered that we have innumerable instances, both in our domestic productions and in those in a state of nature, of all sorts of differences of structure which have become correlated to certain ages, and to either sex. We have differences correlated not only to one sex, but to that short period alone when the reproductive system is active, as in the nuptial plumage of many birds, and in the hooked jaws of the male salmon. We have even slight differences in the horns of different breeds of cattle in relation to an artificially imperfect state of the male sex; for oxen of certain breeds have longer horns than in other breeds, in comparison with the horns of the bulls or cows of these same breeds. Hence I can see no real difficulty in any character having become correlated with the sterile condition of certain members of insect-communities: the difficulty lies in understanding how such correlated modifications of structure could have been slowly accumulated by natural selection.

This difficulty, though appearing insuperable, is lessened, or, as I believe, disappears, when it is remembered that selection may be applied to the family, as well as to the individual, and may thus gain the desired end. Thus, a well-flavoured vegetable is cooked, and the individual is destroyed; but the horticulturist sows seeds of the same stock, and confidently expects to get nearly the same variety; breeders of cattle wish the flesh and fat to be well marbled together; the animal has been slaughtered, but the breeder goes with confidence to the same family. I have such faith in the powers of selection, that I do not doubt that a breed of cattle, always yielding oxen with extraordinarily long horns, could be slowly formed by carefully watching which individual bulls and cows, when matched, produced oxen with the longest horns; and yet no one ox could ever have propagated its kind. Thus I believe it has been with social insects: a slight modification of structure, or instinct, correlated with the sterile condition of certain members of the community, has been advantageous to the community: consequently the fertile males and females of the same community flourished, and transmitted to their fertile offspring a tendency to produce sterile members having the same modification. And I believe that this process has been repeated, until that prodigious amount of difference between the fertile and sterile females of the same species has been produced, which we see in many social insects.

But we have not as yet touched on the climax of the difficulty; namely, the fact that the neuters of several ants differ, not only from the fertile females and males, but from each other, sometimes to an almost incredible degree, and are thus divided into two or even three castes. The castes, moreover, do not generally graduate into each other, but are perfectly well defined; being as distinct from each other, as are any two species of the same genus, or rather as any two genera of the same family. Thus in Eciton, there are working and soldier neuters, with jaws and instincts extraordinarily different: in Cryptocerus, the workers of one caste alone carry a wonderful sort of shield on their heads, the use of which is quite unknown: in the Mexican Myrmecocystus, the workers of one caste never leave the nest; they are fed by the workers of another caste, and they have an enormously developed abdomen which secretes a sort of honey, supplying the place of that excreted by the aphides, or the domestic cattle as they may be called, which our European ants guard or imprison.

It will indeed be thought that I have an overweening confidence in the principle of natural selection, when I do not admit that such wonderful and well-established facts at once annihilate my theory. In the simpler case of neuter insects all of one caste or of the same kind, which have been rendered by natural selection, as I believe to be quite possible, different from the fertile males and females,—in this case, we may safely conclude from the analogy of ordinary variations, that each successive, slight, profitable modification did not probably at first appear in all the individual neuters in the same nest, but in a few alone; and that by the long-continued selection of the fertile parents which produced most neuters with the profitable modification, all the neuters ultimately came to have the desired character. On this view we ought occasionally to find neuter-insects of the same species, in the same nest, presenting gradations of structure; and this we do find, even often, considering how few neuter-insects out of Europe have been carefully examined.

15

Reprinted from *Westminster Rev.* **88:**17-20 (1867)

MIMICRY AND OTHER PROTECTIVE RESEMBLANCES AMONG ANIMALS

A. R. Wallace

[*Editor's Note:* In the original, material precedes this excerpt.]

But there is yet another series of phenomena connected with this subject, which considerably strengthens the view here adopted, while it seems quite incompatible with either of the other hypotheses ; namely, the relation of protective colouring and mimicry to the sexual differences of animals.   It will be clear to every one that if two animals, which as regards " external conditions" and " hereditary descent," are exactly alike, yet differ remarkably in colouration, one resembling a protected species and the other not, the resemblance that exists in one only, can hardly be imputed to the influence of external conditions or as the effect of heredity.   And if, further, it can be proved that the one requires protection more than the other, and that in several cases it is that one which mimics the protected species, while the one that least requires protection never does so, it will afford very strong corroborative evidence that there is a real connexion between the necessity for protection and the phenomenon of mimicry.   Now the sexes of insects offer us a test of the nature here indicated, and appear to furnish one of the most conclusive arguments in favour of the theory that the phenomena termed " mimicry" are produced by natural selection.

The comparative importance of the sexes varies much in different classes of animals.   In the higher vertebrates, where the number of young produced at a birth is small and the same individuals breed many years in succession, the preservation of both sexes is almost equally important.   In all the numerous cases in which the male protects the female and her offspring, or helps to supply them with food, his importance in the economy of nature is proportionately increased, though it is never perhaps quite equal to that of the female.   In insects the case is very different; they pair but once in their lives, and the prolonged existence of the male is in most cases quite unnecessary for the continuance of the race.   The female, however, must continue to exist long enough to deposit her eggs in a place adapted for the development and growth of the progeny.   Hence there is a wide difference in the need for protection in the two sexes: and we should, therefore, expect to find that in some cases the special protection given to the female was in the male less in amount or altogether wanting.   The facts entirely confirm this expectation.   In the spectre insects (Phasmidæ) it is often the females alone that so strikingly resemble leaves, while the males show only a rude approximation.   The male Diadema bolina is a very handsome and conspicuous butterfly, without a sign of protective or imitative colouring, while the female is entirely unlike her partner, and is one of the most wonderful cases of mimicry on record, resembling most accurately the common Danais chrysippus, in whose company it is often found.   So in several species of South American Pieris, the males are white and black, of a similar type of colouring to our own " cabbage" butterflies, while the females are rich yellow and buff, spotted and marked so as exactly to resemble

species of Heliconidæ with which they associate in the forest.   In the Malay archipelago Mr. Wallace found a Diadema which had always been considered a male insect on account of its glossy metallic-blue tints, while its companion of sober brown was looked upon as the female.   He discovered, however, that the reverse is the case, and that the rich and glossy colours of the female are imitative and protective, since they cause her exactly to resemble the common Euplœa midamus of the same regions, a species which has been already mentioned in this article as mimicked by another butterfly, Papilio paradoxa.   In this case, and in that of Diadema bolina, there is no difference in the habits of the two sexes, which fly in similar localities; so that the influence of " external conditions" cannot be invoked here as it has been in the case of the South American Pieris pyrrha and allies, where the white males frequent open sunny places, while the Heliconia-like female haunts the shades of the forest.

We may impute to the same general cause (the greater need of protection for the female, owing to her weaker flight, greater exposure to attack, and supreme importance)—the fact of the colours of female insects being so very generally duller and less conspicuous than those of the other sex.   And that it is chiefly due to this cause rather than to what Mr. Darwin terms "sexual selection" appears to be shown by the otherwise inexplicable fact, that in the groups which have a protection of any kind independent of concealment, sexual differences of colour are either quite wanting or slightly developed.   The Heliconidæ and Danaidæ, protected by a disagreeable flavour, have the females as bright and conspicuous as the males, and very rarely differing at all from them.   The stinging Hymenoptera have the two sexes equally well coloured.   The Carabidæ, the Chrysomelidæ, and the Telephori have both sexes equally conspicuous, and seldom differing in colours.   The brilliant Curculios, which are protected by their hardness, are brilliant in both sexes.   Lastly, the glittering Cetoniadæ and Buprestidæ, which seem to be protected by their hard and polished coats, their rapid motions and peculiar habits, present few sexual differences of colour, while sexual selection has often manifested itself by structural differences, such as horns, spines, or other processes.

The same law manifests itself in Birds.   The female while sitting on her eggs requires protection by concealment to a much greater extent than the male ; and we accordingly find that in a large majority of the cases in which the male birds are distinguished by unusual brilliancy of plumage, the females are much more obscure, and often remarkably plain-coloured.   The exceptions are such as eminently prove the rule, for in most cases we can see a very good reason for them.   In particular, there are a few instances among wading and gallinaceous birds in which the female has decidedly more brilliant colours than the male; but it is a most curious and interesting fact that in most if not all these cases the males sit upon the eggs; so that this exception to the usual rule almost demonstrates that it is because the process of incubation is at once very important and very dangerous, that the protection of obscure colouring is developed. The most striking example is that of the sooty phalarope (Phalaropus fulicarius, Linn.). In winter plumage the sexes of this bird are alike in colouration, but in summer the female is much the most conspicuous, having a black head, dark wings, and reddish-brown back, while the male is nearly uniform brown, with dusky spots. Mr. Gould in his " Birds of Great Britain " figures the two sexes in both winter and summer plumage, and remarks on the strange peculiarity of the usual colours of the two sexes being reversed, and also on the still more curious fact that the " male alone sits on the eggs," which are deposited on the bare ground. In another British bird, the dotterell, the female is also larger and more brightly coloured than the male; and it seems to be proved that the males assist in incubation even if they do not perform it entirely, for Mr. Gould tells us, " that they have been shot with the breast bare of feathers, caused by sitting on the eggs." The small quail-like birds forming the genus Turnix have also generally large and bright-coloured females, and we are told by Mr. Jerdon in his "Birds of India " that " the natives report that during the breeding season the females desert their eggs and associate in flocks while the males are employed in hatching the eggs." It is also an ascertained fact, that the females are more bold and pugnacious than the males. A further confirmation of this view is to be found in the fact (not hitherto noticed) that in a large majority of the cases in which bright colours exist in both sexes incubation takes place in a dark hole or in a dome-shaped nest. Female king-fishers are often equally brilliant with the male, and they build in holes in banks. Bee-eaters, trogons, motmots, and toucans, all build in holes, and in none is there any difference in the sexes, although they are, without exception, showy birds. Parrots build in holes in trees, and in the majority of cases they present no marked sexual difference tending to concealment of the female. Woodpeckers are in the same category, since though the sexes often differ in colour, the female is not generally less conspicuous than the male. Wagtails and titmice build concealed nests, and the females are nearly as gay as their mates. The female of the pretty Australian bird Pardalotus punctatus, is very conspicuously spotted on the upper surface, and it builds in a hole in the ground. The gay-coloured hang-nests (Icterinæ) and the equally brilliant Tanagers may be well contrasted; for the former, concealed in their covered nests, present little or no sexual difference of colour,—while the open-nested Tanagers have the females dull-coloured and sometimes with almost protective tints. No doubt there are many individual exceptions to the rule here indicated, because many and various causes have combined to determine both the colouration and the habits of birds. These have no doubt acted and re-acted on each other; and then under changed conditions it may well have happened that one has become modified, while the other has been continued by hereditary descent, and exists as an apparent exception to what otherwise seems a very general rule. The facts presented to us by the sexual differences of colour in birds and their mode of nesting, are on the whole in perfect harmony with that law of protective adaptation of colour and form, which appears to have checked to some extent the powerful action of sexual selection, and to have materially influenced the colouring of female birds, as it has undoubtedly done that of female insects.

We have now completed a brief, and necessarily very imperfect survey of the various ways in which the external form and colouring of animals is adapted to be useful to them, either by concealing them from their enemies or from the creatures they prey upon. It has, we hope, been shown that the subject is one of much interest, both as regards a true comprehension of the place each animal fills in the economy of nature, and the means by which it is enabled to maintain that place; and also as teaching us how important a part is played by the minutest details in the structure of animals, and how complicated and delicate is the equilibrium of the organic world.

Our exposition of the subject having been necessarily somewhat lengthy and full of details, it will be as well to recapitulate its main points. There is a general harmony in nature between the colours of an animal and those of its habitation. Arctic animals are white, desert animals are sand-coloured, dwellers among leaves and grass are green, nocturnal animals are dusky. These colours are not universal, but are very general, and are seldom reversed. Going on a little further, we find birds, reptiles, and insects tinted and mottled so as exactly to match the rock, or bark, or leaf, or flower they are accustomed to rest upon,—and thereby effectually concealed. Another step in advance, and we have insects which are formed as well as coloured so as exactly to resemble particular leaves, or sticks, or mossy twigs, or flowers; and in these cases very peculiar habits and instincts come into play to aid in the deception and render the concealment more natural. We now enter upon a new phase of the phenomena, and come to creatures whose colours neither conceal them nor make them like vegetable or mineral substances; on the contrary, they are conspicuous enough, but they completely resemble some other creature of a quite different group, while they differ much in outward appearance from those with which all essential parts of their organization show them to be really closely allied. They appear like actors or masqueraders dressed up and painted for amusement, or like swindlers endeavouring to pass themselves off for well-known and respectable members of society. What is the meaning of this strange travestie?

Does Nature descend to imposture or masquerade? We answer, she does not. Her principles are too severe. There is a use in every detail of her handiwork. The resemblance of one animal to another is of exactly the same essential nature as the resemblance to a leaf, or to bark, or to desert sand, and answers exactly the same purpose. In the one case the enemy will not attack the leaf or the bark, and so the disguise is a safeguard; in the other case it is found that for various reasons the creature resembled is passed over and not attacked by the usual enemies of its order, and thus the creature that resembles it has an equally effectual safeguard. We are plainly shown that the disguise is of the same nature in the two cases, by the occurrence in the same group of one species resembling a vegetable substance, while another resembles a living animal of another group; and we know that the creatures resembled possess an immunity from attack, by their being always very abundant, by their being conspicuous and not concealing themselves, and by their having generally no visible means of escape from their enemies; while, at the same time, the particular quality that makes them disliked is often very clear, such as a nasty taste or an indigestible hardness. Further examination reveals the fact that, in several cases of both kinds of disguise, it is the female only that is thus disguised; and as it can be shown that the female needs protection much more than the male, and that her preservation for a much longer period is absolutely necessary for the continuance of the race, we have an additional indication that the resemblance is in all cases subservient to a great purpose—the preservation of the species.

In endeavouring to explain these phenomena as having been brought about by variation and natural selection, we start with the fact that white varieties frequently occur, and when protected from enemies show no incapacity for continued existence and increase. We know, further, that varieties of many other tints occasionally occur; and as "the survival of the fittest" must inevitably weed out those whose colours are prejudicial and preserve those whose colours are a safeguard, we require no other mode of accounting for the protective tints of arctic and desert animals. But this being granted, there is such a perfectly continuous and graduated series of examples of every kind of protective imitation, up to the most wonderful cases of what is termed "mimicry," that we can find no place at which to draw the line, and say, "so far variation and natural selection will account for the phenomena, but for all the rest we require a more potent cause." The counter theories that have been proposed, that of the "special creation" of each imitative form, that of the action of "similar conditions of existence" for some of the cases, and of the laws of "hereditary descent and the reversion to ancestral forms" for others,—have all been shown to be beset with difficulties, and the two latter to be directly contradicted by some of the most constant and most remarkable of the facts to be accounted for.

The important part that "protective resemblance" has played in determining the colours and markings of many groups of animals, will enable us to understand the meaning of one of the most striking facts in nature, the uniformity in the colours of the vegetable as compared with the wonderful diversity of the animal world. There appears no good reason why trees and shrubs should not have been adorned with as many varied hues and as strikingly designed patterns as birds and butterflies, since the gay colours of flowers show that there is no incapacity in vegetable tissues to exhibit them. But even flowers themselves present us with none of those wonderful designs, those complicated arrangements of stripes and dots and patches of colour, that harmonious blending of hues in lines and bands and shaded spots, which are so general a feature in insects. It is the opinion of Mr. Darwin that we owe all the beauty of flowers to the necessity of attracting insects to aid in their fertilization, and that much of the development of colour in the animal world is due to "sexual selection," colour being universally attractive, and thus leading to its propagation and increase; but while fully admitting this, it will be evident, from the facts and arguments here brought forward, that very much of the *variety* both of colour and markings among animals, is due to the supreme importance of concealment, and thus the various tints of minerals and vegetables have been directly reproduced in the animal kingdom, and again and again modified as more special protection became necessary. We shall thus have two causes for the development of colour in the animal world, and shall be better enabled to understand how, by their combined and separate action, the immense variety we now behold has been produced. Both causes, however, will come under the general law of "Utility," the advocacy of which, in its broadest sense, we owe almost entirely to Mr. Darwin.[*] A more accurate knowledge of the varied phenomena connected with this subject may not improbably give us some information both as to the senses and the mental faculties of the lower animals. For it is evident that if colours which please us also attract them, and if the various disguises which have been here enumerated are equally deceptive to them as to ourselves, then both their powers of vision and their faculties of perception and emotion must be essentially of the same nature as our own—a fact of high philosophical importance in the study of our own nature and our true relations to the lower animals.

Although such a variety of interesting facts have been already accumulated, the subject we have been discussing is one of which comparatively little is really known. The natural history of the tropics has never yet been studied on the spot with a full appreciation of "what to observe" in this matter. The varied ways in which the colouring and form of animals serves for their protection, their strange dis-

[*] Mr. Darwin has recognized the fact, that the colouring of female birds has been influenced by the need of protection during incubation. See "Origin of Species," 4th Ed., p. 241.

guises as vegetable or mineral substances, their wonderful mimicry of other beings, offer an almost unworked and inexhaustible field of discovery for the zoologist, and will assuredly throw much light on the laws and conditions which have resulted in the wonderful variety of colour, shade, and marking which constitutes one of the most pleasing characteristics of the animal world, but the immediate causes of which it has hitherto been most difficult to explain.

If we have succeeded in showing that in this wide and picturesque domain of nature, results which have hitherto been supposed to depend either upon those incalculable combinations of laws which we term chance, or upon the direct volition of the Creator, are really due to the action of comparatively well-known and simple causes, we shall have attained our present purpose, which has been to extend the interest so generally felt in the more striking facts of natural history to a large class of curious but much neglected details; and to further, in however slight a degree, the subjection of the phenomena of life to the "Reign of Law."

Part IV

CONTROVERSY OVER *THE DESCENT OF MAN AND SELECTION IN RELATION TO SEX*, 1871–1882

Editor's Comments
on Papers 16 Through 20

Charles Darwin considered sexual selection to be such an important factor in the evolution of human races that he published his thoughts on sexual selection and human evolution in one two-volume book in 1871—*Descent of Man and Selection in Relation to Sex* (excerpts in papers 16A, B, C, and D). Darwin wrote in his introduction that "the sole object of this work is to consider, firstly, whether man, like every other species, is descended from some pre-existing form; secondly, the manner of his development; and thirdly, the value of the differences between the so-called races of man" (Darwin, 1871, vol. 1, pp. 2–3). He also pointed out why he devoted over two-thirds of the book to a discussion of sexual selection:

> During many years it has seemed to me highly probable that sexual selection has played an important part in differentiating the races of man; but in my *Origin of Species* (first edition, p. 199) I contented myself by merely alluding to this belief. When I came to apply this view to man, I found it indispensable to treat the whole subject in full detail. Consequently the second part of the present work, treating of sexual selection, has extended to an inordinate length, compared with the first part; but this could not be avoided (Darwin, 1871, vol. 1, pp. 4–5).

In the chapter "On the Races of Man" with which Darwin concluded the seven chapters of the book that dealt with the "Descent of Man," Darwin wrote:

> We have thus far been baffled in all our attempts to account for the differences between the races of man [direct action of the conditions of life, effects of continued use of parts, the principle of correlated growth, natural selection]; but there remains one important agency, namely Sexual Selection, which appears to have acted as powerfully on man, as on many other animals. (Darwin, 1871, vol. 1, p. 249, reprinted in Paper 16B)

He then devoted the next chapter to the "Principles of Sexual Selection" (excerpts reprinted in Paper 16C), ten chapters to a review of the evidence for sexual selection in nonhuman species, two chapters to sexual selection in relation to man, and one chapter to his "General Summary and Conclusion" on the descent of man and on sexual selection (excerpt reprinted in Paper 16D).

In his chapter "The Principles of Sexual Selection," Darwin recalled why he had originally divided selection operating in nature into natural selection and sexual selection:

> Sexual selection must have come into action, for the males have acquired their present structure, not from being better fitted to survive in the struggle for existence, but from having gained an advantage over other males, and from having transmitted this advantage to their male offspring alone. It was the importance of this distinction which led me to designate this form of selection as sexual selection. (Darwin, 1871, vol. 1, p. 257, reprinted in Paper 16C)

Darwin also distinguished between sexual selection and natural selection:

> Sexual selection acts in a less rigorous manner than natural selection. The latter produces its effects by the life or death at all ages of the more or less successful individuals. Death, indeed, not rarely ensues from the conflicts of rival males. But generally the less successful male merely fails to obtain a female, or obtains later in the season a retarded and less vigorous female, or, if polygamous, obtains fewer females; so that they leave fewer, or

less vigorous, or no offspring. (Darwin, 1871, vol. 1, p. 278, reprinted in Paper 16C)

He then contended that while natural selection sets limits on the adaptations sexual selection can produce:

> The advantages which favoured males have derived from conquering other males in battle or courtship, and thus leaving a more numerous progeny, have been in the long run greater than those derived from rather more perfect adaptation to the external conditions of life. (Darwin, 1871, vol. 1, p. 279, in Paper 16C)

Darwin's strong advocacy of sexual selection theory demonstrates that he realized that selection operates at the level of the individual and can favor individual adaptations that are harmful to other members of the group or even harmful to the species as a whole. The ecological level at which Darwin discussed the operation of selection has been analyzed by Ruse (1980, p. 615) who concluded that "apart from some slight equivocation over man, Darwin opted firmly for hypotheses supposing selection always to work at the level of the individual rather than the group."

Faced with the problem of no experimental evidence for the existence of female choice, Darwin employed the comparative method to determine what characters might have evolved by sexual selection operating via female choice (pp. 257-258 in Paper 16C). The question of whether female choice involved consciousness on the part of animals was discussed by Darwin:

> With respect to female birds feeling a preference for particular males, we must bear in mind that we can judge of choice being exerted, only by placing ourselves in imagination in the same position. . . . Does the male parade his charms with so much pomp and rivalry for no purpose? Are we not justified in believing that the female exerts a choice, and that she receives the addresses of the male who pleases her most? It is not probable that she consciously deliberates; but she is most excited or attracted by the most beautiful, or melodious, or gallant males . . . . We can judge, as already remarked, of choice being exerted, only from the analogy of our own minds; and the mental powers of birds, if reason be excluded, do not differ fundamentally from ours. From these various considerations we may conclude that the pairing of birds is not left to chance; but that those males which are best able by their various charms to please or excite the female, are under ordinary circumstances accepted. If this be admitted, there is not much difficulty in understanding how male birds have gradually acquired their ornamental characters. All animals present individual differences, and as man can modify his domesticated birds by selecting the individuals which appear to him the most beautiful, so the habitual or even occasional preference by the female of the more attractive males would almost certainly lead to their modification; and such modifications might in the

course of time be augmented to almost any extent, compatible with the existence of the species. (Darwin, 1871, vol. 2, pp. 122–124)

The existence of sex-reversed species—species in which the male makes the larger parental investment in rearing offspring—was used by Alfred Wallace (1867, pp. 17–18, reprinted in Paper 15) to contend that natural selection for concealment during paternal as well as maternal care of offspring—rather than sexual selection for bright colors and ornaments that attract the opposite sex—was responsible for much of the sexual dimorphism that exists with respect to coloration and ornamentation in these species. The attributes of sex-reversed species among fishes and birds were summarized by Darwin (1871, vol. 2, pp. 18–22, 200–208, 236), who used them to defend his theory of sexual selection. Darwin noted that in *Dromoeus irroratus* (a species of emu belonging to the ostrich order of birds), where the female is considerably larger than the male and possesses an ornamental top-knot, "we have a complete reversal not only of the parental and incubating instincts, but of the usual moral qualities of the two sexes; the females being savage, quarrelsome and noisy, the males gentle and good" (Darwin, 1871, vol. 2, p. 205). Darwin contended that sexual dimorphism with respect to coloration, ornamentation, and size in sex-reversed species is the result of the direction of sexual selection being reversed due to the replacement of both male-male competition by female-female competition and female choice of mate by male choice.

The relationship that exists between the relative parental investment by the two sexes within a species and their behavior with respect to intrasexual competition for mates and mate choice has been recently reviewed by Trivers (1972), who demonstrates that sexual selection is not only widespread throughout the organic world but is of major importance in the evolution of behavior. Trivers (1972, p. 173) concludes that "The relative parental investment of the sexes in their young is the key variable controlling the operation of sexual selection."

While Darwin considered female choice to be an important force in the evolution of the human species (Darwin, 1871, vol. 2, pp. 368–375, 383), he contended that female choice had become checked in societies by such practices as female infanticide, "early betrothals and the low estimation in which women are held, as mere slaves" (Darwin, 1871, vol. 2, pp. 358–368). Darwin argued that the male "gained the power of selection" during the course of human evolution because "Man is more powerful in body and mind than woman," which enabled him in the savage state to keep the female" in a far more abject state of bondage than does the male of any other animal" (Darwin, 1871, vol. 2, p. 371).

The scenario of the evolution of human intelligence that Charles Darwin (Paper 16A) constructed illustrates the way in which he attempted to relate selection operating at the level of the individual with selection operating at the level of the group. Darwin hypothesized that selection operating at the level of the individual could bring about the advancement of intellectual faculties by favoring the reproduction of the "more sagacious" individual and/or his genetic kin—"blood relatives" or "family." According to Darwin, selection could also favor one tribe (group) over another because the inventions of the more sagacious individuals could lead to the tribe gaining an advantage over other tribes and supplanting them. Group selection operating via wars between tribes was also assigned an important role by Darwin (1871, vol. 1, pp. 162-163, 166) when he hypothesized about the evolution of human social and moral faculties.

Darwin contended that sexual selection operated in monogamous species as well as in polygamous species. He argued that assortative mating should be favored by sexual selection in a monogamous species of birds because the offspring of matings between the most vigorous males and most vigorous females would be more vigorous as well as more numerous (Paper 16C). This particular theory that Darwin championed has been studied by O'Donald (1980, pp. 3, 136-234) who has concluded that: "Earliness of breeding time has been shown to correlate with breeding success in many non-passerine birds producing exactly the conditions that Darwin postulated for sexual selection" among monogamous species; and "computer simulations of Darwin's model show that sexual selection can give rise to large selective coefficients in favour of attractive males whom the females prefer to mate with." Selection for assortative mating has been reviewed by Thiessen and Gregg (1980) and by O'Donald (1980).

Darwin was criticized by Alfred Wallace for trying to explain too much by sexual selection in a book review (Paper 17) of *The Descent of Man and Selection in Relation to Sex* that Wallace wrote at Darwin's request. Wallace did state that sexual selection "must be admitted to have exerted a most powerful influence over the higher forms of life" (p. 177 in Paper 17). He questioned Darwin's contention that female choice played a role in the evolution of sexual dimorphism in insects. Wallace also questioned female choice in general by asking how it was possible for females to share a preference for a particular attribute such as color in the males, and for this shared preference to persist over hundreds of successive generations (p. 182 in Paper 17). Wallace later claimed that it was his belief that these two conditions are not met in nature, which led him to reject female choice as a factor in the evolution of sexual dimorphism in all species (Kottler, 1980). Wallace championed his case against female choice being a force in the evolution of sexual dimorphism in "The Colours of Animals" (Wallace, 1877), which he revised and published in *Tropical Nature and Other Essays* (Wallace, 1878). He published his definitive position opposing female choice in *Darwinism* (Wallace, 1889, excerpts in Paper 22).

St. George Mivart (1830–1900), an English biologist who argued that species evolve by saltations (jumps) directed by some unknown "internal innate force" (Mivart, 1871), wrote a scathing review (Paper 18) of *The Descent of Man and Selection in Relation to Sex*. He contended that " 'Sexual selection' is the corner-stone of Mr. Darwin's theory" and that the consequent "assignment of the law of 'natural selection' to a subordinate position is virtually an abandonment of the Darwinian theory; for the one distinguishing feature of that theory was the all-sufficiency of 'natural selection' " (pp. 48, 53 in Paper 18).

In spite of the fact that Mivart's contention was blatantly false, it has passed into the folklore of science (Hull, 1973, p. 412). Darwin responded to Mivart's attack by arranging and paying for the publication of a reprint of a review by Chauncey Wright attacking the theories Mivart advocated in his book *On the Genesis of Species* (Darwin, 1899, vol. 2, pp. 326–328; Wright, 1871; see also Hull, 1973); by inserting several passages in the last edition of his *Origin of Species by Means of Natural Selection* countering Mivart's objections; and by ending his preface to the second edition of *Descent of Man and Selection in Relation to Sex* with the following statement:

> it has been said by several critics, that when I found that many details of structure in man could not be explained through natural selection, I invented sexual selection; I gave, however, a tolerably clear sketch of this principle in the first edition of the "Origin of Species," and I there stated that it was applicable to man. This subject of sexual selection has been treated at full length in the present work, simply because an opportunity was here first afforded me. I have been struck with the likeness of many of the half-favourable criticisims on sexual selection, with those which appeared at first on natural selection; such as, that it would explain some few details, but certainly was not applicable to the extent to which I have employed it. My conviction of the power of sexual selection remains unshaken; but it is probable, or almost certain, that several of my conclusions will hereafter be found erroneous; this can hardly fail to be the case in the first treatment of a subject. When naturalists have become familiar with the idea of sexual selection, it will, as I believe, be much more largely accepted; and it has already been fully and favourably received by several capable judges. (Darwin, 1874, p. vi)

In the second edition Darwin emphasized his contention that butter-flies engage in a display of beauty and that the taste for the beautiful was permanent enough to allow for the operation of female choice in lower animals (Darwin, 1874, pp. 316–320, 329). Darwin's views concerning the importance of female choice were supported in 1874 by Thomas Belt (Paper 19), who contended that sex-limited mimicry in butterflies could only be explained if sexual selection for the male's ability to successfully court a potential mate operating via female choice was stronger than natural selection for mimicry in males operating via predation. The effectiveness of female choice in producing sex-limited mimicry among butterflies has been recently reviewed by Turner (1978).

Darwin's theory of sexual selection was attacked by Mivart (1876) and Wallace (1877, 1878). Darwin continued to defend his theory, publishing two papers on the subject during the last decade of his life, the first discussing *Sexual Selection in Relation to Monkeys* (Darwin, 1876). In his autobiography written in 1876, Darwin stated that sexual selection was one of the subjects he had "been able to write in full, so as to use all the materials which I had collected" (Barlow, 1958, p. 131). Darwin used a manuscript sent to him during the last months of his life to express his views on the importance of sexual selection one last time. He wrote *A Preliminary Notice* (Paper 20) to *On the Modification of a Race of Syrian Street-Dogs by Means of Sexual Selection* by Dr. Van Dyk. Darwin defended his theory that female animals exert choice with respect to the males they mate with while admitting: "It would, however, be more correct to speak of the females as being excited or attracted in an especial degree by the appearance, voice, &c. of certain males, rather than of deliberately selecting them" (p. 25 in Paper 20).

Darwin emphasized his belief in the importance of sexual selection in this *Preliminary Notice,* (p. 367 in Paper 20), which was read only a few hours before his death at a meeting of the Zoological Society:

> I may perhaps be here permitted to say that, after having carefully weighed, to the best of my ability, the various arguments which have been advanced against the principle of sexual selection, I remain firmly convinced of its truth.

REFERENCES

Barlow, N., ed., 1958, *The Autobiography of Charles Darwin,* Harcourt, Brace and World, New York, 253p. (original omissions restored, appendix and notes added).

Darwin, C. R., 1871, *The Descent of Man and Selection in Relation to Sex,* vols. 1 and 2, J. Murray, London, 423p., 475p. (facsimile of 1st ed., Princeton University Press, 1981).

Darwin, C. R., 1874, *The Descent of Man and Selection in Relation to Sex,* 2nd ed., J. Murray, London, 688p.

Darwin, C. R., 1876, Sexual Selection in Relation to Monkeys, *Nature* **15** (Nov. 2): 18–19. (Reprinted in P. H. Barrett, ed., 1977, *The Collected Papers of Charles Darwin,* vol. 2, University of Chicago Press, Chicago.)

Darwin, F., ed., 1899, *The Life and Letters of Charles Darwin,* 2 vols., Appleton, New York, 558p., 562p.

Darwin, F., and A. C. Seward, eds., 1903, *More Letters of Charles Darwin,* 2 vols., Appleton, New York, 494p., 508p.

Hull, D. L., ed., 1973, *Darwin and His Critics: The Reception of Darwin's Theory of Evolution by the Scientific Community,* Harvard University Press, Cambridge, Mass., 473p.

Kottler, M., 1980, Darwin, Wallace, and the Origin of Sexual Dimorphism, *Am. Philos. Soc. Proc.* **124:**203–226.

Marchant, J., ed., 1916, *Alfred Russel Wallace: Letters and Reminiscences,* Harper, New York, 507p.

Mivart, G., 1871, *On the Genesis of Species,* Appleton, New York, 314p.

Mivart, G., 1876, *Lessons from Nature,* Appleton, New York, 462p.

O'Donald, P., 1980, *Genetic Models of Sexual Selection,* Cambridge University Press, Cambridge, England, 250p.

Peckham, M., ed., 1959, *The Origin of Species by Charles Darwin, A Variorum Text,* University of Pennsylvania Press, Philadelphia, Pa., 816p.

Ruse, M., 1980, Charles Darwin and Group Selection, *Ann. Sci.* **37:**615–630.

Stecher, R. M., 1969, The Darwin-Bates Letters: Correspondence between Two Nineteenth-Century Travellers and Naturalists, Part I, *Ann. Sci.* **25:**1–47.

Thiessen, D., and B. Gregg, 1980, Human Assortative Mating and Genetic Equilibrium: An Evolutionary Perspective, *Ethol. Sociobiol.* **1:**111–140.

Trivers, R., 1972, Parental Investment and Sexual Selection, in *Sexual Selection and the Descent of Man 1871-1971,* B. Campbell, ed., Aldine, Chicago, pp. 136–179.

Turner, J. R. G., 1978, Why Male Butterflies are Non-Mimetic: Natural Selection, Sexual Selection, Group Selection, Modification and Sieving, *Linn. Soc. London (Biol.) J.* **10:**385–432.

Wallace, A. R., 1877, The Colours of Animals and Plants. Part I. The Colours of Animals, *Macmillan's Mag.* **36:**384–408.

Wallace, A. R., 1878, *Tropical Nature and Other Essays,* Macmillan, London, 356p.

Wright, C., 1871, Review of "Genesis of Species," *North Am. Rev.* **13**(July): 63–103. (Reprinted in D. L. Hull, ed., 1973, *Darwin and His Critics,* Harvard University Press, Cambridge, Mass.)

16A

Reprinted from pages 160–161 of *The Descent of Man and Selection in Relation to Sex*, vol. 1, J. Murray, London, 1871, 423p.

ON THE DEVELOPMENT OF THE INTELLECTUAL AND MORAL FACULTIES DURING PRIMEVAL AND CIVILISED TIMES

C. R. Darwin

[*Editor's Note:* In the original, material precedes and follows this excerpt.]

All that we know about savages, or may infer from their traditions and from old monuments, the history of which is quite forgotten by the present inhabitants, shew that from the remotest times successful tribes have supplanted other tribes. Relics of extinct or forgotten tribes have been discovered throughout the civilised regions of the earth, on the wild plains of America, and on the isolated islands in the Pacific Ocean. At the present day civilised nations are everywhere supplanting barbarous nations, excepting where the climate opposes a deadly barrier; and they succeed mainly, though not exclusively, through their arts, which are the products of the intellect. It is, therefore, highly probable that with mankind the intellectual faculties have been gradually perfected through natural selection; and this conclusion is sufficient for our purpose. Undoubtedly it would have been very interesting to have traced the development of each separate faculty from the state in which it exists in the lower animals to that in which it exists in man; but neither my ability nor knowledge permit the attempt.

It deserves notice that as soon as the progenitors of man became social (and this probably occurred at a very early period), the advancement of the intellectual faculties will have been aided and modified in an important manner, of which we see only traces in the lower animals, namely, through the principle of imitation, together with reason and experience. Apes are much given to imitation, as are the lowest savages; and the simple fact previously referred to, that after a time no animal can be caught in the same place by the same sort of trap, shews that animals learn by experience, and imitate each others' caution. Now, if some one man in a tribe, more sagacious than the others, invented a new snare or weapon, or other means of attack or defence, the plainest self-interest, without the assistance of much reasoning power, would prompt the other members to imitate him; and all would thus profit. The habitual practice of each new art must likewise in some slight degree strengthen the intellect. If the new invention were an important one, the tribe would increase in number, spread, and supplant other tribes. In a tribe thus rendered more numerous there would always be a rather better chance of the birth of other superior and inventive members. If such men left children to inherit their mental superiority, the chance of the birth of still more ingenious members would be somewhat better, and in a very small tribe decidedly better. Even if they left no children, the tribe would still include their blood-relations; and it has been ascertained by agriculturists[4] that by preserving and breeding from the family of an animal, which when slaughtered was found to be valuable, the desired character has been obtained.

---

[4] I have given instances in my 'Variation of Animals under Domestication,' vol. ii. p. 196.

16B

Reprinted from pages 248–250 of *The Descent of Man and Selection in Relation to Sex*, vol. 1, J. Murray, London, 1871, 423p.

THE RACES OF MAN

C. R. Darwin

[*Editor's Note:* In the original, material precedes this excerpt.]

We have now seen that the characteristic differences between the races of man cannot be accounted for in a satisfactory manner by the direct action of the conditions of life, nor by the effects of the continued use of parts, nor through the principle of correlation. We are therefore led to inquire whether slight individual differences, to which man is eminently liable, may not have been preserved and augmented during a long series of generations through natural selection. But here we are at once met by the objection that beneficial variations alone can be thus preserved; and as far as we are enabled to judge (although always liable to error on this head) not one of the external differences between the races of man are of any direct or

---

[53] Mr. Catlin states (' N. American Indians,' 3rd edit. 1842, vol. i. p. 49) that in the whole tribe of the Mandans, about one in ten or twelve of the members of all ages and both sexes have bright silvery grey hair, which is hereditary. Now this hair is as coarse and harsh as that of a horse's mane, whilst the hair of other colours is fine and soft.

[54] On the odour of the skin, Godron, 'Sur l'Espèce,' tom. ii. p. 217. On the pores in the skin, Dr. Wilckens, 'Die Aufgaben der landwirth. Zoot.chuik,' 1869, s. 7.

special service to him.   The intellectual and moral or
social faculties must of course be excepted from this remark; but differences in these faculties can have had
little or no influence on external characters.   The variability of all the characteristic differences between the
races, before referred to, likewise indicates that these
differences cannot be of much importance; for, had
they been important, they would long ago have been
either fixed and preserved, or eliminated.   In this
respect man resembles those forms, called by naturalists
protean or polymorphic, which have remained extremely
variable, owing, as it seems, to their variations being of
an indifferent nature, and consequently to their having
escaped the action of natural selection.

We have thus far been baffled in all our attempts
to account for the differences between the races of man;
but there remains one important agency, namely Sexual
Selection, which appears to have acted as powerfully
on man, as on many other animals.   I do not intend
to assert that sexual selection will account for all the
differences between the races.   An unexplained residuum is left, about which we can in our ignorance
only say, that as individuals are continually born with,
for instance, heads a little rounder or narrower, and
with noses a little longer or shorter, such slight differences might become fixed and uniform, if the unknown agencies which induced them were to act in a
more constant manner, aided by long-continued inter-
crossing.   Such modifications come under the provisional class, alluded to in our fourth chapter, which for
the want of a better term have been called spontaneous
variations.   Nor do I pretend that the effects of sexual
selection can be indicated with scientific precision; but
it can be shewn that it would be an inexplicable fact if
man had not been modified by this agency, which has

acted so powerfully on innumerable animals, both high and low in the scale.   It can further be shewn that the differences between the races of man, as in colour, hairyness, form of features, &c., are of the nature which it might have been expected would have been acted on by sexual selection.   But in order to treat this subject in a fitting manner, I have found it necessary to pass the whole animal kingdom in review; I have therefore devoted to it the Second Part of this work.   At the close I shall return to man, and, after attempting to shew how far he has been modified through sexual selection, will give a brief summary of the chapters in this First Part.

16c

Reprinted from pages 253–279 and 296–300 of *The Descent of Man and Selection in Relation to Sex,* vol. 1, J. Murray, London, 1871, 423p.

PRINCIPLES OF SEXUAL SELECTION

C. R. Darwin

Secondary sexual characters — Sexual selection — Manner of action — Excess of males — Polygamy — The male alone generally modified through sexual selection — Eagerness of the male — Variability of the male — Choice exerted by the female — Sexual compared with natural selection — Inheritance, at corresponding periods of life, at corresponding seasons of the year, and as limited by sex — Relations between the several forms of inheritance — Causes why one sex and the young are not modified through sexual selection — Supplement on the proportional numbers of the two sexes throughout the animal kingdom — On the limitation of the numbers of the two sexes through natural selection.

WITH animals which have their sexes separated, the males necessarily differ from the females in their organs of reproduction; and these afford the primary sexual characters. But the sexes often differ in what Hunter has called secondary sexual characters, which are not directly connected with the act of reproduction; for instance, in the male possessing certain organs of sense or locomotion, of which the female is quite destitute, or in having them more highly-developed, in order that he may readily find or reach her; or again, in the male having special organs of prehension so as to hold her securely. These latter organs of infinitely diversified kinds graduate into, and in some cases can hardly be distinguished from, those which are commonly ranked as primary, such as the complex appendages at the apex of the abdomen in male insects. Unless indeed we confine the term "primary" to the reproductive glands, it is scarcely possible to decide, as far as the organs of prehension are concerned, which ought to be called primary and which secondary.

The female often differs from the male in having organs for the nourishment or protection of her young, as the mammary glands of mammals, and the abdominal sacks of the marsupials. The male, also, in some few cases differs from the female in possessing analogous organs, as the receptacles for the ova possessed by the males of certain fishes, and those temporarily developed in certain male frogs. Female bees have a special apparatus for collecting and carrying pollen, and their ovipositor is modified into a sting for the defence of their larvæ and the community. In the females of many insects the ovipositor is modified in the most complex manner for the safe placing of the eggs. Numerous similar cases could be given, but they do not here concern us. There are, however, other sexual differences quite disconnected with the primary organs with which we are more especially concerned— such as the greater size, strength, and pugnacity of the male, his weapons of offence or means of defence against rivals, his gaudy colouring and various ornaments, his power of song, and other such characters.

Besides the foregoing primary and secondary sexual differences, the male and female sometimes differ in structures connected with different habits of life, and not at all, or only indirectly, related to the reproductive functions. Thus the females of certain flies (Culicidæ and Tabanidæ) are blood-suckers, whilst the males live on flowers and have their mouths destitute of mandibles.[1] The males alone of certain moths and of some

---

[1] Westwood, 'Modern Class. of Insects,' vol. ii. 1840, p. 541. In

crustaceans (*e.g.* Tanais) have imperfect, closed mouths, and cannot feed. The Complemental males of certain cirripedes live like epiphytic plants either on the female or hermaphrodite form, and are destitute of a mouth and prehensile limbs. In these cases it is the male which has been modified and has lost certain important organs, which the other members of the same group possess. In other cases it is the female which has lost such parts; for instance, the female glowworm is destitute of wings, as are many female moths, some of which never leave their cocoons. Many female parasitic crustaceans have lost their natatory legs. In some weevil-beetles (Curculionidæ) there is a great difference between the male and female in the length of the rostrum or snout;[2] but the meaning of this and of many analogous differences, is not at all understood. Differences of structure between the two sexes in relation to different habits of life are generally confined to the lower animals; but with some few birds the beak of the male differs from that of the female. No doubt in most, but apparently not in all these cases, the differences are indirectly connected with the propagation of the species: thus a female which has to nourish a multitude of ova will require more food than the male, and consequently will require special means for procuring it. A male animal which lived for a very short time might without detriment lose through disuse its organs for procuring food; but he would retain his locomotive organs in a perfect state, so that he might reach the female. The female, on the other hand, might safely lose her organs for flying, swimming,

---

regard to the statement about Tanais, mentioned below, I am indebted to Fritz Müller.

[2] Kirby and Spence, 'Introduction to Entomology,' vol. iii. 1826, p. 309.

or walking, if she gradually acquired habits which rendered such powers useless.

We are, however, here concerned only with that kind of selection, which I have called sexual selection. This depends on the advantage which certain individuals have over other individuals of the same sex and species, in exclusive relation to reproduction. When the two sexes differ in structure in relation to different habits of life, as in the cases above mentioned, they have no doubt been modified through natural selection, accompanied by inheritance limited to one and the same sex. So again the primary sexual organs, and those for nourishing or protecting the young, come under this same head; for those individuals which generated or nourished their offspring best, would leave, *cæteris paribus*, the greatest number to inherit their superiority; whilst those which generated or nourished their offspring badly, would leave but few to inherit their weaker powers. As the male has to search for the female, he requires for this purpose organs of sense and locomotion, but if these organs are necessary for the other purposes of life, as is generally the case, they will have been developed through natural selection. When the male has found the female he sometimes absolutely requires prehensile organs to hold her; thus Dr. Wallace informs me that the males of certain moths cannot unite with the females if their tarsi or feet are broken. The males of many oceanic crustaceans have their legs and antennæ modified in an extraordinary manner for the prehension of the female; hence we may suspect that owing to these animals being washed about by the waves of the open sea, they absolutely require these organs in order to propagate their kind, and if so their development will have been the result of ordinary or natural selection.

When the two sexes follow exactly the same habits

of life, and the male has more highly developed sense or locomotive organs than the female, it may be that these in their perfected state are indispensable to the male for finding the female; but in the vast majority of cases, they serve only to give one male an advantage over another, for the less well-endowed males, if time were allowed them, would succeed in pairing with the females; and they would in all other respects, judging from the structure of the female, be equally well adapted for their ordinary habits of life. In such cases sexual selection must have come into action, for the males have acquired their present structure, not from being better fitted to survive in the struggle for existence, but from having gained an advantage over other males, and from having transmitted this advantage to their male offspring alone. It was the importance of this distinction which led me to designate this form of selection as sexual selection. So again, if the chief service rendered to the male by his prehensile organs is to prevent the escape of the female before the arrival of other males, or when assaulted by them, these organs will have been perfected through sexual selection, that is by the advantage acquired by certain males over their rivals. But in most cases it is scarcely possible to distinguish between the effects of natural and sexual selection. Whole chapters could easily be filled with details on the differences between the sexes in their sensory, locomotive, and prehensile organs. As, however, these structures are not more interesting than others adapted for the ordinary purposes of life, I shall almost pass them over, giving only a few instances under each class.

There are many other structures and instincts which must have been developed through sexual selection—such as the weapons of offence and the means of defence possessed by the males for fighting with and driving away their rivals—their courage and pugnacity—their ornaments of many kinds—their organs for producing vocal or instrumental music — and their glands for emitting odours; most of these latter structures serving only to allure or excite the female. That these characters are the result of sexual and not of ordinary selection is clear, as unarmed, unornamented, or unattractive males would succeed equally well in the battle for life and in leaving a numerous progeny, if better endowed males were not present. We may infer that this would be the case, for the females, which are unarmed and unornamented, are able to survive and procreate their kind. Secondary sexual characters of the kind just referred to, will be fully discussed in the following chapters, as they are in many respects interesting, but more especially as they depend on the will, choice, and rivalry of the individuals of either sex. When we behold two males fighting for the possession of the female, or several male birds displaying their gorgeous plumage, and performing the strangest antics before an assembled body of females, we cannot doubt that, though led by instinct, they know what they are about, and consciously exert their mental and bodily powers.

In the same manner as man can improve the breed of his game-cocks by the selection of those birds which are victorious in the cockpit, so it appears that the strongest and most vigorous males, or those provided with the best weapons, have prevailed under nature, and have led to the improvement of the natural breed or species. Through repeated deadly contests, a slight degree of variability, if it led to some advantage, however slight, would suffice for the work of sexual selection; and it is certain that secondary sexual characters are eminently variable. In the same manner as man can give beauty, according to his standard of taste, to his male poultry—can give to the Sebright bantam a new and elegant plumage, an erect and peculiar carriage—so it appears that in a state of nature female birds, by having long selected the more attractive males, have added to their beauty. No doubt this implies powers of discrimination and taste on the part of the female which will at first appear extremely improbable; but I hope hereafter to shew that this is not the case.

From our ignorance on several points, the precise manner in which sexual selection acts is to a certain extent uncertain. Nevertheless if those naturalists who already believe in the mutability of species, will read the following chapters, they will, I think, agree with me that sexual selection has played an important part in the history of the organic world. It is certain that with almost all animals there is a struggle between the males for the possession of the female. This fact is so notorious that it would be superfluous to give instances. Hence the females, supposing that their mental capacity sufficed for the exertion of a choice, could select one out of several males. But in numerous cases it appears as if it had been specially arranged that there should be a struggle between many males. Thus with migratory birds, the males generally arrive before the females at their place of breeding, so that many males are ready to contend for each female. The bird-catchers assert that this is invariably the case with the nightingale and blackcap, as I am informed by Mr. Jenner Weir, who confirms the statement with respect to the latter species.

Mr. Swaysland of Brighton, who has been in the habit, during the last forty years, of catching our migratory birds on their first arrival, writes to me that he has never known the females of any species to arrive before their males. During one spring he shot thirty-nine males of Ray's wagtail (*Budytes Raii*) before he saw a single female. Mr. Gould has ascertained by dissection, as he informs me, that male snipes arrive in this country before the females. In the case of fish, at the period when the salmon ascend our rivers, the males in large numbers are ready to breed before the females. So it apparently is with frogs and toads. Throughout the great class of insects the males almost always emerge from the pupal state before the other sex, so that they generally swarm for a time before any females can be seen.[3] The cause of this difference between the males and females in their periods of arrival and maturity is sufficiently obvious. Those males which annually first migrated into any country, or which in the spring were first ready to breed, or were the most eager, would leave the largest number of offspring; and these would tend to inherit similar instincts and constitutions. On the whole there can be no doubt that with almost all animals, in which the sexes are separate, there is a constantly recurrent struggle between the males for the possession of the females.

Our difficulty in regard to sexual selection lies in understanding how it is that the males which conquer other males, or those which prove the most attractive to the females, leave a greater number of offspring to inherit their superiority than the beaten and less

---

[3] Even with those of plants in which the sexes are separate, the male flowers are generally mature before the female. Many hermaphrodite plants are, as first shewn by C. K. Sprengel, dichogamous; that is, their male and female organs are not ready at the same time, so that they cannot be self-fertilised. Now with such plants the pollen is generally mature in the same flower before the stigma, though there are some exceptional species in which the female organs are mature before the male.

attractive males.   Unless this result followed, the characters which gave to certain males an advantage over others, could not be perfected and augmented through sexual selection.   When the sexes exist in exactly equal numbers, the worst-endowed males will ultimately find females (excepting where polygamy prevails), and leave as many offspring, equally well fitted for their general habits of life, as the best-endowed males.   From various facts and considerations, I formerly inferred that with most animals, in which secondary sexual characters were well developed, the males considerably exceeded the females in number; and this does hold good in some few cases.   If the males were to the females as two to one, or as three to two, or even in a somewhat lower ratio, the whole affair would be simple; for the better-armed or more attractive males would leave the largest number of offspring.   But after investigating, as far as possible, the numerical proportions of the sexes, I do not believe that any great inequality in number commonly exists.   In most cases sexual selection appears to have been effective in the following manner.

Let us take any species, a bird for instance, and divide the females inhabiting a district into two equal bodies: the one consisting of the more vigorous and better-nourished individuals, and the other of the less vigorous and healthy.   The former, there can be little doubt, would be ready to breed in the spring before the others; and this is the opinion of Mr. Jenner Weir, who has during many years carefully attended to the habits of birds.   There can also be no doubt that the most vigorous, healthy, and best-nourished females would on an average succeed in rearing the largest number of offspring.   The males, as we have seen, are generally ready to breed before the females; of the males the strongest, and with some species the best armed, drive away the weaker males; and the former would then unite with the more vigorous and best-nourished females, as these are the first to breed. Such vigorous pairs would surely rear a larger number of offspring than the retarded females, which would be compelled, supposing the sexes to be numerically equal, to unite with the conquered and less powerful males; and this is all that is wanted to add, in the course of successive generations, to the size, strength and courage of the males, or to improve their weapons.

But in a multitude of cases the males which conquer other males, do not obtain possession of the females, independently of choice on the part of the latter. The courtship of animals is by no means so simple and short an affair as might be thought. The females are most excited by, or prefer pairing with, the more ornamented males, or those which are the best songsters, or play the best antics; but it is obviously probable, as has been actually observed in some cases, that they would at the same time prefer the more vigorous and lively males.[4] Thus the more vigorous females, which are the first to breed, will have the choice of many males; and though they may not always select the strongest or best armed, they will select those which are vigorous and well armed, and in other respects the most attractive. Such early pairs would have the same advantage in rearing offspring on the female side as above explained, and nearly the same advantage on the male side. And this apparently has sufficed during a long course of generations to add not only to the strength and fighting-powers of

---

[4] I have received information, hereafter to be given, to this effect with respect to poultry. Even with birds, such as pigeons, which pair for life, the female, as I hear from Mr. Jenner Weir, will desert her mate if he is injured or grows weak.

the males, but likewise to their various ornaments or other attractions.

In the converse and much rarer case of the males selecting particular females, it is plain that those which were the most vigorous and had conquered others, would have the freest choice; and it is almost certain that they would select vigorous as well as attractive females. Such pairs would have an advantage in rearing offspring, more especially if the male had the power to defend the female during the pairing-season, as occurs with some of the higher animals, or aided in providing for the young. The same principles would apply if both sexes mutually preferred and selected certain individuals of the opposite sex; supposing that they selected not only the more attractive, but likewise the more vigorous individuals.

Numerical Proportion of the Two Sexes.—I have remarked that sexual selection would be a simple affair if the males considerably exceeded in number the females. Hence I was led to investigate, as far as I could, the proportions between the two sexes of as many animals as possible; but the materials are scanty. I will here give only a brief abstract of the results, retaining the details for a supplementary discussion, so as not to interfere with the course of my argument. Domesticated animals alone afford the opportunity of ascertaining the proportional numbers at birth; but no records have been specially kept for this purpose. By indirect means, however, I have collected a considerable body of statistical data, from which it appears that with most of our domestic animals the sexes are nearly equal at birth. Thus with race-horses, 25,560 births have been recorded during twenty-one years, and the male births have been to the female births as 99·7 to 100. With greyhounds the inequality is greater than with any other animal, for during twelve years, out of 6878 births, the male births have been as 110·1 to 100 female births. It is, however, in some degree doubtful whether it is safe to infer that the same proportional numbers would hold good under natural conditions as under domestication; for slight and unknown differences in the conditions affect to a certain extent the proportion of the sexes. Thus with mankind, the male births in England are as 104·5, in Russia as 108·9, and with the Jews of Livornia as 120 to 100 females. The proportion is also mysteriously affected by the circumstance of the births being legitimate or illegitimate.

For our present purpose we are concerned with the proportion of the sexes, not at birth, but at maturity, and this adds another element of doubt; for it is a well ascertained fact that with man a considerably larger proportion of males than of females die before or during birth, and during the first few years of infancy. So it almost certainly is with male lambs, and so it may be with the males of other animals. The males of some animals kill each other by fighting; or they drive each other about until they become greatly emaciated. They must, also, whilst wandering about in eager search for the females, be often exposed to various dangers. With many kinds of fish the males are much smaller than the females, and they are believed often to be devoured by the latter or by other fishes. With some birds the females appear to die in larger proportion than the males: they are also liable to be destroyed on their nests, or whilst in charge of their young. With insects the female larvæ are often larger than those of the males, and would consequently be more likely to be devoured: in some cases the mature females are less active and less rapid in their movements than the males, and would not be so well able to escape from danger. Hence, with animals in a state of nature, in order to judge of the proportions of the sexes at maturity, we must rely on mere estimation; and this, except perhaps when the inequality is strongly marked, is but little trustworthy.   Nevertheless, as far as a judgment can be formed, we may conclude from the facts given in the supplement, that the males of some few mammals, of many birds, of some fish and insects, considerably exceed in number the females.

The proportion between the sexes fluctuates slightly during successive years: thus with race-horses, for every 100 females born, the males varied from 107·1 in one year to 92·6 in another year, and with greyhounds from 116·3 to 95·3.   But had larger numbers been tabulated throughout a more extensive area than England, these fluctuations would probably have disappeared; and such as they are, they would hardly suffice to lead under a state of nature to the effective action of sexual selection.   Nevertheless with some few wild animals, the proportions seem, as shewn in the supplement, to fluctuate either during different seasons or in different localities in a sufficient degree to lead to such action. For it should be observed that any advantage gained during certain years or in certain localities by those males which were able to conquer other males, or were the most attractive to the females, would probably be transmitted to the offspring and would not subsequently be eliminated.   During the succeeding seasons, when from the equality of the sexes every male was everywhere able to procure a female, the stronger or more attractive males previously produced would still have at least as good a chance of leaving offspring as the less strong or less attractive.

Polygamy.—The practice of polygamy leads to the

same results as would follow from an actual inequality
in the number of the sexes; for if each male secures
two or more females, many males will not be able to
pair; and the latter assuredly will be the weaker or
less attractive individuals.   Many mammals and some
few birds are polygamous, but with animals belonging to
the lower classes I have found no evidence of this habit.
The intellectual powers of such animals are, perhaps,
not sufficient to lead them to collect and guard a harem
of females.   That some relation exists between polygamy and the development of secondary sexual characters, appears nearly certain; and this supports the
view that a numerical preponderance of males would
be eminently favourable to the action of sexual selection.
Nevertheless many animals, especially birds, which are
strictly monogamous, display strongly-marked secondary
sexual characters; whilst some few animals, which are
polygamous, are not thus characterised.

We will first briefly run through the class of mammals, and then turn to birds.   The gorilla seems to be
a polygamist, and the male differs considerably from
the female; so it is with some baboons which live in
herds containing twice as many adult females as males.
In South America the *Mycetes caraya* presents well-
marked sexual differences in colour, beard, and vocal
organs, and the male generally lives with two or three
wives: the male of the *Cebus capucinus* differs somewhat from the female, and appears to be polygamous.[5]
Little is known on this head with respect to most other
monkeys, but some species are strictly monogamous.
The ruminants are eminently polygamous, and they

---

[5] On the Gorilla, Savage and Wyman, 'Boston Journal of Nat. Hist.'
vol. v. 1845-47, p. 423.   On Cynocephalus, Brehm, 'Illust. Thierleben,'
B. i. 1864, s. 77.   On Mycetes, Rengger, 'Naturgesch.: Säugethiere
von Paraguay,' 1830, s. 14, 20.   On Cebus, Brehm, ibid. s. 108.

more frequently present sexual differences than almost any other group of mammals, especially in their weapons, but likewise in other characters.   Most deer, cattle, and sheep are polygamous; as are most antelopes, though some of the latter are monogamous.   Sir Andrew Smith, in speaking of the antelopes of South Africa, says that in herds of about a dozen there was rarely more than one mature male.   The Asiatic *Antilope saiga* appears to be the most inordinate polygamist in the world; for Pallas[6] states that the male drives away all rivals, and collects a herd of about a hundred, consisting of females and kids: the female is hornless and has softer hair, but does not otherwise differ much from the male.   The horse is polygamous, but, except in his greater size and in the proportions of his body, differs but little from the mare.   The wild boar, in his great tusks and some other characters, presents well-marked sexual characters; in Europe and in India he leads a solitary life, except during the breeding-season; but at this season he consorts in India with several females, as Sir W. Elliot, who has had large experience in observing this animal, believes: whether this holds good in Europe is doubtful, but is supported by some statements.   The adult male Indian elephant, like the boar, passes much of his time in solitude; but when associating with others, "it is rare to find," as Dr. Campbell states, "more than one male with a whole " herd of females."   The larger males expel or kill the smaller and weaker ones.   The male differs from the female by his immense tusks and greater size, strength, and endurance; so great is the difference in these latter

---

[6] Pallas, 'Spicilegia Zoolog.' Fasc. xii. 1777, p. 29.   Sir Andrew Smith, 'Illustrations of the Zoology of S. Africa,' 1849, pl. 29, on the Kobus.   Owen, in his 'Anatomy of Vertebrates' (vol. iii. 1868, p. 633) gives a table incidentally showing which species of Antelopes pair and which are gregarious.

respects, that the males when caught are valued at twenty per cent. above the females.[7] With other pachydermatous animals the sexes differ very little or not at all, and they are not, as far as known, polygamists. Hardly a single species amongst the Cheiroptera and Edentata, or in the great Orders of the Rodents and Insectivora, presents well-developed secondary sexual differences; and I can find no account of any species being polygamous, excepting, perhaps, the common rat, the males of which, as some rat-catchers affirm, live with several females.

The lion in South Africa, as I hear from Sir Andrew Smith, sometimes lives with a single female, but generally with more than one, and, in one case, was found with as many as five females, so that he is polygamous. He is, as far as I can discover, the sole polygamist in the whole group of the terrestrial Carnivora, and he alone presents well-marked sexual characters. If, however, we turn to the marine Carnivora, the case is widely different; for many species of seals offer, as we shall hereafter see, extraordinary sexual differences, and they are eminently polygamous. Thus the male sea-elephant of the Southern Ocean, always possesses, according to Péron, several females, and the sea-lion of Forster is said to be surrounded by from twenty to thirty females. In the North, the male sea-bear of Steller is accompanied by even a greater number of females.

With respect to birds, many species, the sexes of which differ greatly from each other, are certainly monogamous. In Great Britain we see well-marked sexual differences in, for instance, the wild-duck which pairs with a single female, with the common blackbird,

---

[7] Dr. Campbell, in 'Proc. Zoolog. Soc.' 1869, p. 138. See also an interesting paper, by Lieut. Johnstone, in 'Proc. Asiatic Soc. of Bengal,' May, 1868.

and with the bullfinch which is said to pair for life.   So
it is, as I am informed by Mr. Wallace, with the Chat-
terers or Cotingidæ of South America, and numerous
other birds.   In several groups I have not been able to
discover whether the species are polygamous or monogamous.   Lesson says that birds of paradise, so remarkable for their sexual differences, are polygamous,
but Mr. Wallace doubts whether he had sufficient evidence.   Mr. Salvin informs me that he has been led
to believe that humming-birds are polygamous.   The
male widow-bird, remarkable for his caudal plumes,
certainly seems to be a polygamist.[8]   I have been
assured by Mr. Jenner Weir and by others, that three
starlings not rarely frequent the same nest; but whether
this is a case of polygamy or polyandry has not been
ascertained.

The Gallinaceæ present almost as strongly marked
sexual differences as birds of paradise or humming-
birds, and many of the species are, as is well known,
polygamous; others being strictly monogamous.   What
a contrast is presented between the sexes by the polygamous peacock or pheasant, and the monogamous
guinea-fowl or partridge!   Many similar cases could
be given, as in the grouse tribe, in which the males
of the polygamous capercailzie and black-cock differ
greatly from the females; whilst the sexes of the monogamous red grouse and ptarmigan differ very little.
Amongst the Cursores, no great number of species
offer strongly - marked sexual differences, except the
bustards, and the great bustard (*Otis tarda*), is said to

---

[8] 'The Ibis,' vol. iii. 1861, p. 133, on the Progne Widow-bird. See
also on the Vidua axillaris, ibid. vol. ii. 1860, p. 211.   On the polygamy of the Capercailzie and Great Bustard, see L. Lloyd, 'Game Birds
of Sweden,' 1867, p. 19, and 182.   Montagu and Selby speak of the
Black Grouse as polygamous and of the Red Grouse as monogamous.

be polygamous. With the Grallatores, extremely few species differ sexually, but the ruff (*Machetes pugnax*) affords a strong exception, and this species is believed by Montagu to be a polygamist. Hence it appears that with birds there often exists a close relation between polygamy and the development of strongly-marked sexual differences. On asking Mr. Bartlett, at the Zoological Gardens, who has had such large experience with birds, whether the male tragopan (one of the Gallinaceæ) was polygamous, I was struck by his answering, "I do not know, but should think so from "his splendid colours."

It deserves notice that the instinct of pairing with a single female is easily lost under domestication. The wild-duck is strictly monogamous, the domestic-duck highly polygamous. The Rev. W. D. Fox informs me that with some half-tamed wild-ducks, kept on a large pond in his neighbourhood, so many mallards were shot by the gamekeeper that only one was left for every seven or eight females; yet unusually large broods were reared. The guinea-fowl is strictly monogamous; but Mr. Fox finds that his birds succeed best when he keeps one cock to two or three hens.[9] Canary-birds pair in a state of nature, but the breeders in England successfully put one male to four or five females; nevertheless the first female, as Mr. Fox has been assured, is alone treated as the wife, she and her young ones being fed by him; the others are treated as concubines. I have noticed these cases, as it renders it in some degree probable that monogamous species, in a state of nature, might readily become either temporarily or permanently polygamous.

---

[9] The Rev. E. S. Dixon, however, speaks positively ('Ornamental Poultry,' 1848, p. 76) about the eggs of the guinea-fowl being infertile when more than one female is kept with the same male.

With respect to reptiles and fishes, too little is known of their habits to enable us to speak of their marriage arrangements. The stickle-back Gasterosteus), however, is said to be a polygamist;[10] and the male during the breeding-season differs conspicuously from the female.

To sum up on the means through which, as far as we can judge, sexual selection has led to the development of secondary sexual characters. It has been shewn that the largest number of vigorous offspring will be reared from the pairing of the strongest and best-armed males, which have conquered other males, with the most vigorous and best-nourished females, which are the first to breed in the spring. Such females, if they select the more attractive, and at the same time vigorous, males, will rear a larger number of offspring than the retarded females, which must pair with the less vigorous and less attractive males. So it will be if the more vigorous males select the more attractive and at the same time healthy and vigorous females; and this will especially hold good if the male defends the female, and aids in providing food for the young. The advantage thus gained by the more vigorous pairs in rearing a larger number of offspring has apparently sufficed to render sexual selection efficient. But a large preponderance in number of the males over the females would be still more efficient; whether the preponderance was only occasional and local, or permanent; whether it occurred at birth, or subsequently from the greater destruction of the females; or whether it indirectly followed from the practice of polygamy.

The Male generally more modified than the Female.—Throughout the animal kingdom, when the sexes differ

---

[10] Noel Humphreys, 'River Gardens,' 1857.

from each other in external appearance, it is the male which, with rare exceptions, has been chiefly modified; for the female still remains more like the young of her own species, and more like the other members of the same group.   The cause of this seems to lie in the males of almost all animals having stronger passions than the females.   Hence it is the males that fight together and sedulously display their charms before the females; and those which are victorious transmit their superiority to their male offspring.   Why the males do not transmit their characters to both sexes will hereafter be considered.   That the males of all mammals eagerly pursue the females is notorious to every one.   So it is with birds; but many male birds do not so much pursue the female, as display their plumage, perform strange antics, and pour forth their song, in her presence.   With the few fish which have been observed, the male seems much more eager than the female; and so it is with alligators, and apparently with Batrachians.   Throughout the enormous class of insects, as Kirby remarks,[11] "the law is, that the male " shall seek the female."   With spiders and crustaceans, as I hear from two great authorities, Mr. Blackwall and Mr. C. Spence Bate, the males are more active and more erratic in their habits than the females.   With insects and crustaceans, when the organs of sense or locomotion are present in the one sex and absent in the other, or when, as is frequently the case, they are more highly developed in the one than the other, it is almost invariably the male, as far as I can discover, which retains such organs, or has them most developed; and this shews that the male is the more active member in the courtship of the sexes.[12]

---

[11] Kirby and Spence, 'Introduction to Entomology,' vol. iii. 1826, p. 342.

[12] One parasitic Hymenopterous insect (Westwood, ' Modern Class. of Insects,' vol. ii. p. 160) forms an exception to the rule, as the male

The female, on the other hand, with the rarest exception, is less eager than the male. As the illustrious Hunter[13] long ago observed, she generally "requires to "be courted;" she is coy, and may often be seen endeavouring for a long time to escape from the male. Every one who has attended to the habits of animals will be able to call to mind instances of this kind. Judging from various facts, hereafter to be given, and from the results which may fairly be attributed to sexual selection, the female, though comparatively passive, generally exerts some choice and accepts one male in preference to others. Or she may accept, as appearances would sometimes lead us to believe, not the male which is the most attractive to her, but the one which is the least distasteful. The exertion of some choice on the part of the female seems almost as general a law as the eagerness of the male.

We are naturally led to enquire why the male in so many and such widely distinct classes has been rendered more eager than the female, so that he searches for her and plays the more active part in courtship. It would be no advantage and some loss of power if both sexes were mutually to search for each other; but why should the male almost always be the seeker? With plants, the ovules after fertilisation have to be nourished for a time; hence the pollen is necessarily brought to the female organs—being placed on the stigma, through the agency of insects or of the wind,

---

has rudimentary wings, and never quits the cell in which it is born, whilst the female has well-developed wings. Audouin believes that the females are impregnated by the males which are born in the same cells with them; but it is much more probable that the females visit other cells, and thus avoid close interbreeding. We shall hereafter meet with a few exceptional cases, in various classes, in which the female, instead of the male, is the seeker and wooer.

[13] 'Essays and Observations,' edited by Owen, vol. i. 1861, p. 194.

or by the spontaneous movements of the stamens; and with the Algæ, &c., by the locomotive power of the antherozooids. With lowly-organised animals permanently affixed to the same spot and having their sexes separate, the male element is invariably brought to the female; and we can see the reason; for the ova, even if detached before being fertilised and not requiring subsequent nourishment or protection, would be, from their larger relative size, less easily transported than the male element. Hence plants [14] and many of the lower animals are, in this respect, analogous. In the case of animals not affixed to the same spot, but enclosed within a shell with no power of protruding any part of their bodies, and in the case of animals having little power of locomotion, the males must trust the fertilising element to the risk of at least a short transit through the waters of the sea. It would, therefore, be a great advantage to such animals, as their organisation became perfected, if the males when ready to emit the fertilising element, were to acquire the habit of approaching the female as closely as possible. The males of various lowly-organised animals having thus aboriginally acquired the habit of approaching and seeking the females, the same habit would naturally be transmitted to their more highly developed male descendants; and in order that they should become efficient seekers, they would have to be endowed with strong passions. The acquirement of such passions would naturally follow from the more eager males leaving a larger number of offspring than the less eager.

The great eagerness of the male has thus indirectly

---

[14] Prof. Sachs ('Lehrbuch der Botanik,' 1870, s. 633) in speaking of the male and female reproductive cells, remarks, "verhält sich die eine "bei der Vereinigung activ, . . . die andere erscheint bei der Verein-"igung passiv."

led to the much more frequent development of secondary sexual characters in the male than in the female. But the development of such characters will have been much aided, if the conclusion at which I arrived after studying domesticated animals, can be trusted, namely, that the male is more liable to vary than the female. I am aware how difficult it is to verify a conclusion of this kind. Some slight evidence, however, can be gained by comparing the two sexes in mankind, as man has been more carefully observed than any other animal. During the Novara Expedition [15] a vast number of measurements of various parts of the body in different races were made, and the men were found in almost every case to present a greater range of variation than the women; but I shall have to recur to this subject in a future chapter. Mr. J. Wood,[16] who has carefully attended to the variation of the muscles in man, puts in italics the conclusion that " the greatest number of " abnormalities in each subject is found in the males." He had previously remarked that " altogether in 102 " subjects the varieties of redundancy were found to " be half as many again as in females, contrasting " widely with the greater frequency of deficiency in " females before described." Professor Macalister likewise remarks[17] that variations in the muscles " are " probably more common in males than females." Certain muscles which are not normally present in mankind are also more frequently developed in the male than in the female sex, although exceptions to this rule

---

[15] 'Reise der Novara: Anthropolog. Theil,' 1867, s. 216-269. The results were calculated by Dr. Weisbach from measurements made by Drs. K. Scherzer and Schwarz. On the greater variability of the males of domesticated animals, see my ' Variation of Animals and Plants under Domestication,' vol. ii. 1868, p. 75.

[16] 'Proceedings Royal Soc.' vol. xvi. July, 1868, p. 519 and 524.

[17] 'Proc. Royal Irish Academy,' vol. x. 1868, p. 123.

are said to occur. Dr. Burt Wilder[18] has tabulated the cases of 152 individuals with supernumerary digits, of which 86 were males, and 39, or less than half, females; the remaining 27 being of unknown sex. It should not, however, be overlooked that women would more frequently endeavour to conceal a deformity of this kind than men. Whether the large proportional number of deaths of the male offspring of man and apparently of sheep, compared with the female offspring, before, during, and shortly after birth (see supplement), has any relation to a stronger tendency in the organs of the male to vary and thus to become abnormal in structure or function, I will not pretend to conjecture.

In various classes of animals a few exceptional cases occur, in which the female instead of the male has acquired well pronounced secondary sexual characters, such as brighter colours, greater size, strength, or pugnacity. With birds, as we shall hereafter see, there has sometimes been a complete transposition of the ordinary characters proper to each sex; the females having become the more eager in courtship, the males remaining comparatively passive, but apparently selecting, as we may infer from the results, the more attractive females. Certain female birds have thus been rendered more highly coloured or otherwise ornamented, as well as more powerful and pugnacious than the males, these characters being transmitted to the female offspring alone.

It may be suggested that in some cases a double process of selection has been carried on; the males having selected the more attractive females, and the latter the more attractive males. This process however, though it might lead to the modification of both sexes,

---

[18] 'Massachusetts Medical Soc.' vol. ii. No. 3, 1868, p. 9.

would not make the one sex different from the other,
unless indeed their taste for the beautiful differed; but
this is a supposition too improbable in the case of any
animal, excepting man, to be worth considering.   There
are, however, many animals, in which the sexes resemble
each other, both being furnished with the same ornaments, which analogy would lead us to attribute to the
agency of sexual selection.   In such cases it may be
suggested with more plausibility, that there has been a
double or mutual process of sexual selection; the more
vigorous and precocious females having selected the
more attractive and vigorous males, the latter having
rejected all except the more attractive females.   But
from what we know of the habits of animals, this view
is hardly probable, the male being generally eager to
pair with any female.   It is more probable that the
ornaments common to both sexes were acquired by one
sex, generally the male, and then transmitted to the offspring of both sexes.   If, indeed, during a lengthened
period the males of any species were greatly to exceed
the females in number, and then during another
lengthened period under different conditions the reverse
were to occur, a double, but not simultaneous, process
of sexual selection might easily be carried on, by which
the two sexes might be rendered widely different.

We shall hereafter see that many animals exist, of
which neither sex is brilliantly coloured or provided
with special ornaments, and yet the members of both
sexes or of one alone have probably been modified
through sexual selection.   The absence of bright tints
or other ornaments may be the result of variations of
the right kind never having occurred, or of the animals
themselves preferring simple colours, such as plain black
or white.   Obscure colours have often been acquired
through natural selection for the sake of protection, and

the acquirement through sexual selection of conspicuous colours, may have been checked from the danger thus incurred. But in other cases the males have probably struggled together during long ages, through brute force, or by the display of their charms, or by both means combined, and yet no effect will have been produced unless a larger number of offspring were left by the more successful males to inherit their superiority, than by the less successful males; and this, as previously shewn, depends on various complex contingencies.

Sexual selection acts in a less rigorous manner than natural selection. The latter produces its effects by the life or death at all ages of the more or less successful individuals. Death, indeed, not rarely ensues from the conflicts of rival males. But generally the less successful male merely fails to obtain a female, or obtains later in the season a retarded and less vigorous female, or, if polygamous, obtains fewer females; so that they leave fewer, or less vigorous, or no offspring. In regard to structures acquired through ordinary or natural selection, there is in most cases, as long as the conditions of life remain the same, a limit to the amount of advantageous modification in relation to certain special ends; but in regard to structures adapted to make one male victorious over another, either in fighting or in charming the female, there is no definite limit to the amount of advantageous modification; so that as long as the proper variations arise the work of sexual selection will go on. This circumstance may partly account for the frequent and extraordinary amount of variability presented by secondary sexual characters. Nevertheless, natural selection will determine that characters of this kind shall not be acquired by the victorious males, which would be injurious to them in any high degree, either by expending too much of their vital powers, or by exposing them to any great danger. The development, however, of certain structures—of the horns, for instance, in certain stags—has been carried to a wonderful extreme; and in some instances to an extreme which, as far as the general conditions of life are concerned, must be slightly injurious to the male. From this fact we learn that the advantages which favoured males have derived from conquering other males in battle or courtship, and thus leaving a numerous progeny, have been in the long run greater than those derived from rather more perfect adaptation to the external conditions of life. We shall further see, and this could never have been anticipated, that the power to charm the female has been in some few instances more important than the power to conquer other males in battle.

[*Editor's Note:* Material has been omitted at this point.]

Summary and concluding remarks.—From the foregoing discussion on the various laws of inheritance, we
learn that characters often or even generally tend to
become developed in the same sex, at the same age,
and periodically at the same season of the year, in
which they first appeared in the parents.   But these
laws, from unknown causes, are very liable to change.
Hence the successive steps in the modification of a
species might readily be transmitted in different ways;
some of the steps being transmitted to one sex,
and some to both; some to the offspring at one age,
and some at all ages.   Not only are the laws of inheritance extremely complex, but so are the causes which
induce and govern variability.   The variations thus
caused are preserved and accumulated by sexual selection, which is in itself an extremely complex affair,
depending, as it does, on ardour in love, courage, and
the rivalry of the males, and on the powers of perception, taste, and will of the female.   Sexual selection will
also be dominated by natural selection for the general
welfare of the species.   Hence the manner in which the
individuals of either sex or of both sexes are affected
through sexual selection cannot fail to be complex in
the highest degree.

When variations occur late in life in one sex, and are

transmitted to the same sex at the same age, the other sex and the young are necessarily left unmodified. When they occur late in life, but are transmitted to both sexes at the same age, the young alone are left unmodified. Variations, however, may occur at any period of life in one sex or in both, and be transmitted to both sexes at all ages, and then all the individuals of the species will be similarly modified. In the following chapters it will be seen that all these cases frequently occur under nature.

Sexual selection can never act on any animal whilst young, before the age for reproduction has arrived. From the great eagerness of the male it has generally acted on this sex and not on the females. The males have thus become provided with weapons for fighting with their rivals, or with organs for discovering and securely holding the female, or for exciting and charming her. When the sexes differ in these respects, it is also, as we have seen, an extremely general law that the adult male differs more or less from the young male; and we may conclude from this fact that the successive variations, by which the adult male became modified, cannot have occurred much before the age for reproduction. How then are we to account for this general and remarkable coincidence between the period of variability and that of sexual selection,—principles which are quite independent of each other? I think we can see the cause: it is not that the males have never varied at an early age, but that such variations have commonly been lost, whilst those occurring at a later age have been preserved.

All animals produce more offspring than can survive to maturity; and we have every reason to believe that death falls heavily on the weak and inexperienced young. If then a certain proportion of the offspring were to vary at birth or soon afterwards, in some manner which at this age was of no service to them, the chance of the preservation of such variations would be small. We have good evidence under domestication how soon variations of all kinds are lost, if not selected. But variations which occurred at or near maturity, and which were of immediate service to either sex, would probably be preserved; as would similar variations occurring at an earlier period in any individuals which happened to survive. As this principle has an important bearing on sexual selection, it may be advisable to give an imaginary illustration. We will take a pair of animals, neither very fertile nor the reverse, and assume that after arriving at maturity they live on an average for five years, producing each year five young. They would thus produce 25 offspring; and it would not, I think, be an unfair estimate to assume that 18 or 20 out of the 25 would perish before maturity, whilst still young and inexperienced; the remaining seven or five sufficing to keep up the stock of mature individuals. If so, we can see that variations which occurred during youth, for instance in brightness, and which were not of the least service to the young, would run a good chance of being utterly lost. Whilst similar variations, which occurring at or near maturity in the comparatively few individuals surviving to this age, and which immediately gave an advantage to certain males, by rendering them more attractive to the females, would be likely to be preserved. No doubt some of the variations in brightness which occurred at an earlier age would by chance be preserved, and eventually give to the male the same advantage as those which appeared later; and this will account for the young males commonly partaking to a certain extent (as may be observed with many birds) of the bright colours of their adult male parents. If only a few of the successive variations in brightness were to occur at a late age, the adult male would be only a little brighter than the young male; and such cases are common.

In this illustration I have assumed that the young varied in a manner which was of no service to them; but many characters proper to the adult male would be actually injurious to the young,—as bright colours from making them conspicuous, or horns of large size from expending much vital force. Such variations in the young would promptly be eliminated through natural selection. With the adult and experienced males, on the other hand, the advantage thus derived in their rivalry with other males would often more than counterbalance exposure to some degree of danger. Thus we can understand how it is that variations which must originally have appeared rather late in life have alone or in chief part been preserved for the development of secondary sexual characters; and the remarkable coincidence between the periods of variability and of sexual selection is intelligible.

As variations which give to the male an advantage in fighting with other males, or in finding, securing, or charming the female, would be of no use to the female, they will not have been preserved in this sex either during youth or maturity. Consequently such variations would be extremely liable to be lost; and the female, as far as these characters are concerned, would be left unmodified, excepting in so far as she may have received them by transference from the male. No doubt if the female varied and transferred serviceable characters to her male offspring, these would be favoured through sexual selection; and then both sexes would thus far be modified in the same manner. But I shall hereafter have to recur to these more intricate contingencies.

In the following chapters, I shall treat of the secondary sexual characters in animals of all classes, and shall endeavour in each case to apply the principles explained in the present chapter. The lowest classes will detain us for a very short time, but the higher animals, especially birds, must be treated at considerable length. It should be borne in mind that for reasons already assigned, I intend to give only a few illustrative instances of the innumerable structures by the aid of which the male finds the female, or, when found, holds her. On the other hand, all structures and instincts by which the male conquers other males, and by which he allures or excites the female, will be fully discussed, as these are in many ways the most interesting.

[*Editor's Note:* Material has been omitted at this point.]

16D

Reprinted from pages 396–405 of *The Descent of Man and Selection in Relation to Sex,* vol. 2, J. Murray, London, 1871, 405p.

GENERAL SUMMARY AND CONCLUDING REMARKS

C. R. Darwin

[*Editor's Note:* In the original, material precedes this excerpt.]

Sexual selection has been treated at great length in these volumes; for, as I have attempted to shew, it has played an important part in the history of the organic world. As summaries have been given to each chapter, it would be superfluous here to add a detailed summary. I am aware that much remains doubtful, but I have endeavoured to give a fair view of the whole case. In the lower divisions of the animal kingdom, sexual selection seems to have done nothing: such animals are often affixed for life to the same spot, or have the two sexes combined in the same individual, or what is still more important, their perceptive and intellectual faculties are not sufficiently advanced to allow of the feelings of love and jealousy, or of the exertion of choice. When, however, we come to the Arthropoda and Vertebrata, even to the lowest classes in these two great Sub-Kingdoms, sexual selection has effected much; and it deserves notice that we here find the intellectual faculties developed, but in two very distinct lines, to the highest standard, namely in the Hymenoptera (ants, bees, &c.) amongst the Arthropoda, and in the Mammalia, including man, amongst the Vertebrata.

In the most distinct classes of the animal kingdom,

with mammals, birds, reptiles, fishes, insects, and even crustaceans, the differences between the sexes follow almost exactly the same rules.   The males are almost always the wooers; and they alone are armed with special weapons for fighting with their rivals.   They are generally stronger and larger than the females, and are endowed with the requisite qualities of courage and pugnacity.   They are provided, either exclusively or in a much higher degree than the females, with organs for producing vocal or instrumental music, and with odoriferous glands.   They are ornamented with infinitely diversified appendages, and with the most brilliant or conspicuous colours, often arranged in elegant patterns, whilst the females are left unadorned.   When the sexes differ in more important structures, it is the male which is provided with special sense-organs for discovering the female, with locomotive organs for reaching her, and often with prehensile organs for holding her.   These various structures for securing or charming the female are often developed in the male during only part of the year, namely the breeding season.   They have in many cases been transferred in a greater or less degree to the females; and in the latter case they appear in her as mere rudiments.   They are lost by the males after emasculation.   Generally they are not developed in the male during early youth, but appear a short time before the age for reproduction.   Hence in most cases the young of both sexes resemble each other; and the female resembles her young offspring throughout life.   In almost every great class a few anomalous cases occur in which there has been an almost complete transposition of the characters proper to the two sexes; the females assuming characters which properly belong to the males.   This surprising uniformity in the laws regulating the differences between the sexes in so many and such widely separated classes, is intelligible if we admit the action throughout all the higher divisions of the animal kingdom of one common cause, namely sexual selection.

Sexual selection depends on the success of certain individuals over others of the same sex in relation to the propagation of the species; whilst natural selection depends on the success of both sexes, at all ages, in relation to the general conditions of life. The sexual struggle is of two kinds; in the one it is between the individuals of the same sex, generally the male sex, in order to drive away or kill their rivals, the females remaining passive; whilst in the other, the struggle is likewise between the individuals of the same sex, in order to excite or charm those of the opposite sex, generally the females, which no longer remain passive, but select the more agreeable partners. This latter kind of selection is closely analogous to that which man unintentionally, yet effectually, brings to bear on his domesticated productions, when he continues for a long time choosing the most pleasing or useful individuals, without any wish to modify the breed.

The laws of inheritance determine whether characters gained through sexual selection by either sex shall be transmitted to the same sex, or to both sexes; as well as the age at which they shall be developed. It appears that variations which arise late in life are commonly transmitted to one and the same sex. Variability is the necessary basis for the action of selection, and is wholly independent of it. It follows from this, that variations of the same general nature have often been taken advantage of and accumulated through sexual selection in relation to the propagation of the species, and through natural selection in relation to the general purposes of life. Hence secondary sexual characters, when equally transmitted to both sexes can be distinguished from ordinary specific characters only by the light of analogy. The modifications acquired through sexual selection are often so strongly pronounced that the two sexes have frequently been ranked as distinct species, or even as distinct genera. Such strongly-marked differences must be in some manner highly important; and we know that they have been acquired in some instances at the cost not only of inconvenience, but of exposure to actual danger.

The belief in the power of sexual selection rests chiefly on the following considerations. The characters which we have the best reason for supposing to have been thus acquired are confined to one sex; and this alone renders it probable that they are in some way connected with the act of reproduction. These characters in innumerable instances are fully developed only at maturity; and often during only a part of the year, which is always the breeding-season. The males (passing over a few exceptional cases) are the most active in courtship; they are the best armed, and are rendered the most attractive in various ways. It is to be especially observed that the males display their attractions with elaborate care in the presence of the females; and that they rarely or never display them excepting during the season of love. It is incredible that all this display should be purposeless. Lastly we have distinct evidence with some quadrupeds and birds that the individuals of the one sex are capable of feeling a strong antipathy or preference for certain individuals of the opposite sex.

Bearing these facts in mind, and not forgetting the marked results of man's unconscious selection, it seems to me almost certain that if the individuals of one sex were during a long series of generations to prefer pairing with certain individuals of the other sex, characterised in some peculiar manner, the offspring would slowly but surely become modified in this same manner. I have not attempted to conceal that, excepting when the males are more numerous than the females, or when polygamy prevails, it is doubtful how the more attractive males succeed in leaving a larger number of offspring to inherit their superiority in ornaments or other charms than the less attractive males; but I have shewn that this would probably follow from the females,—especially the more vigorous females which would be the first to breed, preferring not only the more attractive but at the same time the more vigorous and victorious males.

Although we have some positive evidence that birds appreciate bright and beautiful objects, as with the Bower-birds of Australia, and although they certainly appreciate the power of song, yet I fully admit that it is an astonishing fact that the females of many birds and some mammals should be endowed with sufficient taste for what has apparently been effected through sexual selection; and this is even more astonishing in the case of reptiles, fish, and insects. But we really know very little about the minds of the lower animals. It cannot be supposed that male Birds of Paradise or Peacocks, for instance, should take so much pains in erecting, spreading, and vibrating their beautiful plumes before the females for no purpose. We should remember the fact given on excellent authority in a former chapter, namely that several peahens, when debarred from an admired male, remained widows during a whole season rather than pair with another bird.

Nevertheless I know of no fact in natural history more wonderful than that the female Argus pheasant should be able to appreciate the exquisite shading of the ball-and-socket ornaments and the elegant patterns on the wing-feathers of the male. He who thinks that the male was created as he now exists must admit that the great plumes, which prevent the wings from being used for flight, and which, as well as the primary feathers, are displayed in a manner quite peculiar to this one species during the act of courtship, and at no other time, were given to him as an ornament. If so, he must likewise admit that the female was created and endowed with the capacity of appreciating such ornaments. I differ only in the conviction that the male Argus pheasant acquired his beauty gradually, through the females having preferred during many generations the more highly ornamented males; the æsthetic capacity of the females having been advanced through exercise or habit in the same manner as our own taste is gradually improved. In the male, through the fortunate chance of a few feathers not having been modified, we can distinctly see how simple spots with a little fulvous shading on one side might have been developed by small and graduated steps into the wonderful ball-and-socket ornaments; and it is probable that they were actually thus developed.

Everyone who admits the principle of evolution, and yet feels great difficulty in admitting that female mammals, birds, reptiles, and fish, could have acquired the high standard of taste which is implied by the beauty of the males, and which generally coincides with our own standard, should reflect that in each member of the vertebrate series the nerve-cells of the brain are the direct offshoots of those possessed by the common progenitor of the whole group. It thus becomes intelligible that the brain and mental faculties should be capable under similar conditions of nearly the same course of development, and consequently of performing nearly the same functions.

The reader who has taken the trouble to go through the several chapters devoted to sexual selection, will be able to judge how far the conclusions at which I have arrived are supported by sufficient evidence. If he accepts these conclusions, he may, I think, safely extend them to mankind; but it would be superfluous here to repeat what I have so lately said on the manner in which sexual selection has apparently acted on both the male and female side, causing the two sexes of man to differ in body and mind, and the several races to differ from each other in various characters, as well as from their ancient and lowly-organised progenitors.

He who admits the principle of sexual selection will be led to the remarkable conclusion that the cerebral system not only regulates most of the existing functions of the body, but has indirectly influenced the progressive development of various bodily structures and of certain mental qualities. Courage, pugnacity, perseverance, strength and size of body, weapons of all kinds, musical organs, both vocal and instrumental, bright colours, stripes and marks, and ornamental appendages, have all been indirectly gained by the one sex or the other, through the influence of love and jealousy, through the appreciation of the beautiful in sound, colour or form, and through the exertion of a choice; and these powers of the mind manifestly depend on the development of the cerebral system.

Man scans with scrupulous care the character and pedigree of his horses, cattle, and dogs before he matches them; but when he comes to his own marriage he rarely, or never, takes any such care. He is impelled by nearly the same motives as are the lower animals when left to their own free choice, though he is in so far superior to them that he highly values mental charms and virtues.   On the other hand he is strongly attracted by mere wealth or rank.   Yet he might by selection do something not only for the bodily constitution and frame of his offspring, but for their intellectual and moral qualities.   Both sexes ought to refrain from marriage if in any marked degree inferior in body or mind; but such hopes are Utopian and will never be even partially realised until the laws of inheritance are thoroughly known.   All do good service who aid towards this end.   When the principles of breeding and of inheritance are better understood, we shall not hear ignorant members of our legislature rejecting with scorn a plan for ascertaining by an easy method whether or not consanguineous marriages are injurious to man.

The advancement of the welfare of mankind is a most intricate problem : all ought to refrain from marriage who cannot avoid abject poverty for their children ; for poverty is not only a great evil, but tends to its own increase by leading to recklessness in marriage.   On the other hand, as Mr. Galton has remarked, if the prudent avoid marriage, whilst the reckless marry, the inferior members will tend to supplant the better members of society.   Man, like every other animal, has no doubt advanced to his present high condition through a struggle for existence consequent on his rapid multiplication; and if he is to advance still higher he must remain subject to a severe struggle.   Otherwise he would soon sink into indolence, and the more highly-gifted men would not be more successful in the battle of life than the less gifted.   Hence our natural rate of increase, though leading to many and obvious evils, must not be greatly diminished by any means.   There should be open competition for all men ; and the most able should not be prevented by laws or customs from succeeding best and rearing the largest number of offspring.   Important as the struggle for existence has been and even still is, yet as far as the highest part of man's nature is concerned there are other agencies more important. For the moral qualities are advanced, either directly or indirectly, much more through the effects of habit, the reasoning powers, instruction, religion, &c., than through natural selection; though to this latter agency the social instincts, which afforded the basis for the development of the moral sense, may be safely attributed.

The main conclusion arrived at in this work, namely that man is descended from some lowly-organised form, will, I regret to think, be highly distasteful to many persons. But there can hardly be a doubt that we are descended from barbarians. The astonishment which I felt on first seeing a party of Fuegians on a wild and broken shore will never be forgotten by me, for the reflection at once rushed into my mind— such were our ancestors. These men were absolutely naked and bedaubed with paint, their long hair was tangled, their mouths frothed with excitement, and their expression was wild, startled, and distrustful. They possessed hardly any arts, and like wild animals lived on what they could catch; they had no government, and were merciless to every one not of their own small tribe. He who has seen a savage in his native land will not feel much shame, if forced to acknowledge that the blood of some more humble creature flows in his veins. For my own part I would as soon be descended from that heroic little monkey, who braved his dreaded enemy in order to save the life of his keeper; or from that old baboon, who, descending from the mountains, carried away in triumph his young comrade from a crowd of astonished dogs—as from a savage who delights to torture his enemies, offers up bloody sacrifices, practises infanticide without remorse, treats his wives like slaves, knows no decency, and is haunted by the grossest superstitions.

Man may be excused for feeling some pride at having risen, though not through his own exertions, to the very summit of the organic scale; and the fact of his having thus risen, instead of having been aboriginally placed there, may give him hopes for a still higher destiny in the distant future. But we are not here concerned with hopes or fears, only with the truth as far as our reason allows us to discover it. I have given the evidence to the best of my ability; and we must acknowledge, as it seems to me, that man with all his noble qualities, with sympathy which feels for the most debased, with benevolence which extends not only to other men but to the humblest living creature, with his god-like intellect which has penetrated into the movements and constitution of the solar system—with all these exalted powers—Man still bears in his bodily frame the indelible stamp of his lowly origin.

17

Reprinted from *Academy (London)* **2**:177-183 (1871)

A REVIEW AND CRITICISM OF MR. DARWIN'S
DESCENT OF MAN

A. R. Wallace

MR. DARWIN's reputation already stands so high that it may seem difficult to add to it. Yet this work will undoubtedly do so, and will prove almost equally attractive to the naturalist and the general reader. The two large volumes on *Domesticated Animals and Plants* caused some little disappointment to those who looked for easy scientific reading; but the present work will have no such drawback. It is throughout written in the author's clearest style, it is not overloaded with detail, it abounds in curious facts and acute reasoning, and it treats of two great subjects of the very highest interest—the nature and origin of man, and the overwhelming importance of sexual influences in moulding and beautifying the animal world.

The few passages devoted to sexual selection in the *Origin of Species*, led many persons to suppose that it was but a vague hypothesis almost unsupported by direct evidence; and most of its opponents have shown an utter ignorance of, or disbelief in, the whole matter. It will now be seen on what a solid foundation of fact the theory of sexual selection is founded, and how true, as regards this part of his subject at all events, was Mr. Darwin's assertion, that his first volume contained but a mere abstract of the evidence before him, and that he could not be properly judged till the whole mass of facts he had collected were made public. From the reticence with which the sexual relations of animals have been treated in popular works, most of the readers of this book will be astonished to find that a new and inner world of animal life exists, of which they had hitherto had no conception; and that a considerable portion of the form and structure, the weapons, the ornaments, and the colouring of animals, owes its very existence to the separation of the sexes. This new branch of natural history is one of the most striking creations of Mr. Darwin's genius, and it is all his own; and although we believe he imputes far too much to its operation, it must be admitted to have exerted a most powerful influence over the higher forms of life. In the first part of this article we propose to sketch in outline the main facts and arguments adduced, and shall afterwards discuss certain points which seem open to criticism.

Mr. Darwin tells us that he has for many years collected the materials on which this work is mainly founded, without any intention of publishing them, as he did not wish to prejudice the reception of the general doctrine of natural selection. That doctrine has, however, made such rapid and unexpected progress that no danger of this kind any longer exists; and he has therefore put together his materials relating to the origin of man from a lower animal form. Believing that sexual selection has played an important part in differentiating the races of man, he has found it necessary to treat this subject in great detail, which has much increased the bulk of the work.

The first chapter discusses the evidence for the descent of man from some lower form. Not only is man's whole structure comparable, bone by bone and muscle by muscle, with that of other vertebrata, but his close relation to them is shown in a variety of unexpected ways. He is able to receive some animal diseases, as glanders, hydrophobia, &c., showing a close similarity to other animals in blood and tissues. The internal and external parasites of man are of the same families and genera as those of the lower animals. The embryonic development of man is exactly similar to that of other vertebrates, so that at an early period his embryo can hardly be distinguished from theirs; and arteries running in arch-like branches as if to carry blood to branchiæ which are not present in the higher animals, show his affinity to the lower aquatic forms. A little later, the great toe is found standing out from the side of the foot, as it does in the quadrumana. Numerous rudiments occur in man of structures characteristic of lower forms. Many muscles regularly present in apes and other quadrupeds occasionally appear in man. The upper part of the infolded lobe of the ear often presents a pointed projection, the rudiment of the pointed and erectile ears of most mammals. The supra-condyloid foramen, through which the great nerve of the fore limb passes in quadrumana and carnivora, is absent in man; but it occasionally reappears, with the nerve passing through it; and a careful examination of the remains of prehistoric races shows, that this form was more frequent in ancient times than now.

The mental powers of man are then compared with those of the lower animals, and it is shown that the latter possess the rudiments of them all. The origin of the moral sense is next treated of; and although. such eminent writers as Mill, Bain, Herbert Spencer, and Sir John Lubbock, have all given their independent theories on this subject, Mr. Darwin has hit upon a perfectly original view, which is perhaps more satisfactory than any which have preceded it. He maintains that the moral sense arises from the social instincts combined with an active intellect. As soon as the mental faculties became well developed, images of past actions and motives would be incessantly passing through the mind of each individual, and a feeling of dissatisfaction would arise whenever it was perceived that the ever present social instinct had yielded to some other instinct stronger at the time but less enduring. For example, such instincts as hunger, lust, or the desire of vengeance, are immensely strong but are not enduring, and do not leave vivid and easily recalled impressions at all proportionate to their intensity at the time. The feeling of sympathy, the need of companionship, the desire for the approbation of our fellows are, on the other hand, ever present with us, and anything which interfered with these would be a constant source of dissatisfaction. If then a being with a suffi-ciently active mind to recall past actions and see the effects they have produced were, under the impulse of any of the stronger instincts, to rob, starve, kill or injure those who were necessary to the satisfaction of his social instincts, he would inevitably feel dissatisfied with himself at having allowed his passion for a temporary enjoyment, the force of which he could not realise afterwards, to interfere with the satisfaction of his less intense but more permanent desires and instincts. A repetition of such experiences would lead to the feeling that the one kind of instincts was less impor-tant to his welfare than the other. He would class them as passions to be regulated and controlled; and when in spite of his determination to control them he had not done so, he would almost despise himself—would feel remorse—would be rebuked by his conscience. Mr. Darwin shows at some length, that the rudiments of all these instincts and emotions exist in animals; and he argues that the acqui-sition of speech would greatly increase their power; for when each member of the community could express his feelings and wishes, the opinion of his fellows would go to increase the regret felt at having allowed the temporary to overcome the permanent instinct. The effect of this at first, would be to limit "virtue" to that which was for the benefit of the tribe exclusively. Murder, robbery, and treachery within the limits of the tribe would be infamous, but beyond these limits might be even praiseworthy. Thus infanticide is so often not looked upon as a crime, because it is sup-posed to be beneficial to the tribe; and no pity has been felt for the sufferings of enemies, of slaves, or even of women. Owing to its great utility to the tribe, courage is always looked upon as the highest virtue; and for the same reason fidelity and self-sacrifice are always highly esteemed. But intemperance and licentiousness are never counted as vices, because they do not immediately concern any one but the individual and his family. Mr. Darwin concludes, that the moral sense is fundamentally identical with the social instincts, and has been developed for the general good of the community, rather than for its greatest happiness. "General good" is defined as "the means by which the greatest possible number of individuals can be reared in full vigour and health, with all their faculties perfect under the conditions to which they are exposed;" and it is quite conceivable that this may not be always iden-tical with "greatest happiness." If so, the present theory will be a step in advance in the history of the utilitarian philosophy.

The manner of development of man from some lower form is next very fully discussed. The extreme variability of every part of man's bodily structure and mental faculties is shown; the effect of changed conditions whether of locality or of habits is proved to be considerable; and arrested developments, reversions, and correlated variations are all shown to obtain in man exactly as they do in the lower animals. Natural selection must have acted on man, because he multiplies rapidly beyond the means of sub-sistence, because he varies, and because he is exposed to varying external conditions; but Mr. Darwin adopts the views of the present writer, that as soon as man's mind had become moderately developed, the action of natural selection would have been checked, as regards his general structure, and transferred to his mental faculties. It is argued that the advance from animal to man must have taken place before the dispersal of the race over the world; and that in some warm country as large as Australia, New Guinea, or Borneo, "the competition between tribe and tribe would have been sufficient under favourable conditions to have raised man through survival of the fittest, combined with the inherited effects of habit, to his present high position in the organic scale." A separate chapter is devoted to the development of man's intellect, and to the effects of natural selection on civilised nations; and though many of the arguments used are open to criticism, the subject is most interesting, and is discussed with Mr. Darwin's usual clearness and candour.

The next two chapters discuss,—the special affinities of man to certain lower animals, by means of which the line of his genealogy can be traced, and the place and time of his origin approximately determined,—and the nature and probable origin of the several races of man. This last he believes cannot be fully explained without the agency of sexual selection, and this leads to the second part of the work, which treats of sexual differences, their causes and effects, throughout the animal kingdom, in order that the principles deduced from this extensive survey may be applied to explain certain residual phenomena in man.

The subject of sexual selection, occupying nearly five hundred pages, is treated in great detail, and abounds in matter of interest; but only a very brief sketch can here be given of it. The main theory depends upon the fact that there is almost invariably a struggle among the males for the females; a struggle carried on either by actual fighting or by rivalry in voice or in beauty. This struggle is moreover ensured by the circumstance that in most cases the males are ready to breed before the females, male insects emerging sooner from the pupa, and male migratory birds arriving earlier than those of the other sex. From this it results that some males gain the victory over others, and succeed in pairing earlier and with the earliest and most vigorous females. The males are always the most eager, the females generally coy; and Mr. Darwin believes that in almost all cases the female exerts a choice, and rejects those males who please her least. Hence have arisen two sets of modifications in male animals : 1. Weapons of various kinds have been developed, owing to those best able to fight having most frequently left progeny to inherit their superiority ; 2. Musical organs, bright colours, or ornamental appendages, have been developed, through the females preferring those so gifted or adorned. The laws of inheritance are first discussed ; the transmission of characters to the male alone through the female, and the transmission of variations at certain ages to the offspring at the same age, and to one or both sexes. A large portion of the animal kingdom is then passed in review, as respects the differentiation of the sexes and the means by which such differentiation has been produced. This part of the work is illustrated by numerous woodcuts, showing the extraordinary differences of form and structure between the sexes. Many parts of the body have been modified to enable the male to seize and hold the female ; and this is adduced as an argument that the female exerts a choice, and has the power of rejecting any particular male. But this hardly seems to follow, for it may well be maintained that when the more active male seizes a female, she cannot escape, and that she has no means of rejecting him and practically never does so.

The males of a considerable number of homopterous and orthopterous insects emit musical sounds by means of very curious and varied apparatus, and there is no doubt that these sounds serve to attract or charm the female. Among most insects the males fight, but neither spiders nor dragonflies have been observed to do so. Among all other insects than lepidoptera, the sexes are as a rule coloured alike or nearly alike, the exceptions being comparatively few ; but among butterflies especially, diversity of colour is the rule, the males being almost always most brilliantly or most intensely coloured ; and the difference is often so great that the two sexes look like widely different species. Beetles differ more in form than in colour, the males often possessing wonderful horns, spines, or protuberances, immensely long legs or antennæ, or enormous jaws, while in colour they hardly differ at all or are only somewhat brighter. Passing on to the vertebrates, we find that male fishes often fight, and exhibit as much ardour as terrestrial animals ; some of them undergo strange changes of form at the breeding season, and some few differ conspicuously from the females in colour, or by the possession of elongated fins, spines, or other appendages. In other cases, although the sexes are usually alike, yet in the breeding season the males acquire new or more vivid colours.

Passing by amphibians and reptiles, among which many curious sexual characters occur, we come to birds, a class which exhibits them in their highest perfection, and which has furnished Mr. Darwin with the most powerful arguments for the complete development of his theory of sexual selection. Almost every imaginable kind of sexual ornament is here to be found. In an immense number of cases male birds are far more beautifully coloured than the females ; and besides this, they often possess the most gorgeous developments of ornamental plumage, as in the train of the peacock, the wings of the African night-jar, the tail of the lyre-bird and of the resplendent trogon, the crest of the umbrella-bird, and the breast plumes of the bird-of-paradise. Spurs are also developed upon the legs or the wings, and the male is generally larger, and has a louder or more melodious voice. Among birds is found the first direct proof that the female notices and admires increased brilliancy or beauty of colour, or any novel ornament ; and, what is more important, that she exercises choice, rejecting one suitor and choosing another. There is abundant evidence too that the male fully displays all his charms before the females, and some of the facts adduced on this head are most curious and interesting. Mr. Darwin also devotes himself to showing how some of the most marvellous developments of beauty in plumage may have been produced by the constant selection of slight modifications ; and he explains in this manner the origin of the eyed train of the peacock, and the wonderfully decorated wings of the Argus pheasant, with an acuteness and success hardly inferior to that which he exhibited when investigating the structure of coral reefs or of orchids. The four chapters on birds would alone demand a lengthy article to do them justice, but as we shall have to return to this subject when we come to criticise some portion of the theory, it will be as well now to pass over the two chapters on the sexual differences and weapons of the mammalia, and devote some little space to a sketch of the concluding chapters, which again treat of man.

The sexual differences of man are stated to be greater than in most species of quadrumana, while in their general features and mode of development man agrees remarkably with those animals, as an example of which we may quote, that whenever the beard differs in colour from the hair on the head, it is always lighter, both in man and monkeys. The law of battle for wives still prevails among some savages, and to this circumstance Mr. Darwin thinks may be traced the undoubted inferiority of woman, not only in bodily strength but also in courage and perseverance, qualities equally necessary to ensure victory. He thinks also that but for the fortunate circumstance that the law of equal transmission of characters to both sexes has commonly prevailed among mammalia, man might have become as much superior to woman in mind as the peacock surpasses the peahen in plumage. Considerable space is devoted to prove that savages think much of personal appearance, admire certain types of form and complexion, and that probably selection of wives and husbands has been an important agent in determining both the racial and the sexual differences of mankind. The evidence adduced, however, seems only to show that the men as a rule ornament themselves more than the women, and that they do so to be admired by their fellow-men quite as much as by the women ; and also that men of each race admire all the characteristic features of their own race, and abhor any wide departure from it ; the natural effect of which would be to keep the race true, not to favour the production of new races. It is admitted that promiscuous intercourse and infanticide would to a great extent prevent the action of sexual selection ; but it would also be rendered nugatory by the fact that among savages no woman remains unmarried, youth and health being amply sufficient charms to procure her a husband. It also seems very uncertain whether any effect would be produced by the more powerful men possessing themselves of a number of the most beautiful women, and rearing on an average a greater number of children, as Mr. Darwin thinks they would do. Where polygamy prevails the number of children to one father may be very large, but will the number to each mother be as large as with the remainder of the tribe who are forced to practise monogamy ? This important point is not alluded to. The absence of hair on the body is admitted to be a character that cannot be accounted for by "natural selection," because it cannot be conceived to have been a beneficial variation ; but "sexual selection" is supposed to account for it. At an exceedingly early period in our history our semi-human ancestors were hairy, and it is thought that one or both sexes preferred less hair ; and any partial nudity that appeared led to a more early or a more constant wedlock, and thus gave an advantage to such individuals and their more numerous progeny. The example of monkeys and apes is adduced, many of which have bare skin on the face or on other parts of the body ; and the New Zealand proverb, " There is no woman for a hairy man," is thought to bear upon the question. This explanation is by no means satisfactory. The analogy of the quadrumana and of other animals would have some force if there were still hairy and hairless or partially hairless men,—with bare faces and breasts, but hairy backs, for example ; but we have to deal with a complete nudity, which has no parallel in the animal kingdom except in cases where " natural selection " has evidently come into play. That a smooth-skinned race like the New Zealanders should object to hairiness is natural ; for, as Mr. Darwin says, each race admires its own characteristics carried to a moderate extreme. Hairy races would therefore admire abundant hairiness, just as bearded races now admire fine beards ; and any admiration of deficient hairiness would probably be as rare and abnormal as the admiration for partial baldness or scanty hair in women, would be among ourselves. Any individual fancy for such an abnormal peculiarity as deficient hair in a hair-covered animal could produce no effect ; and that any such fancy should become general with our semi-human ancestors, and so produce universal nakedness, does not seem at all probable, when we have no evidence of such a result of sexual selection elsewhere in the whole animal kingdom. It is true, that in that early state the struggle for existence would have been severe, and only the best endowed would have survived ; but unless we suppose a universal and simultaneous fancy among all the most vigorous and therefore probably the most hairy men for what would be then an unnatural character—deficiency of hair in women—and that this fancy should have persisted in all its force for a long series of generations, it is not easy to see how this severe struggle for existence and survival of the fittest would in any way aid sexual selection in abolishing the hairy covering. On the contrary it seems more likely that it would entirely prevent it. We can hardly therefore impute much influence to sexual selection in the case of man, even as regards less important characters than the loss of hair, because it requires the very same tastes to persist in the majority of the race during a period of long and unknown duration. All analogy teaches us that there would be no such identity of taste in successive generations ; and this seems a fatal objection to the belief that any fixed and definite characters could have been produced in man by sexual selection alone.

In his last chapter, Mr. Darwin gives an able summary of the whole argument ; and, while regretting that the result he has arrived at will be highly distasteful to many persons, maintains, that the whole evidence leads to the conclusion that man, notwithstanding his noble qualities and his god-like intellect, still bears in his bodily frame the indelible stamp of his lowly origin.

Having thus sketched in outline the theories advanced by our author, and given a summary of the facts by which he supports them, we have now to notice in more detail certain portions of the argument which appear to rest upon an insecure foundation either of logic or of fact.

The first and most obvious objection that will be made to this great work is, that it consists of two books mixed together. The whole of the matter relating to sexual selection among animals, would have formed a fitting third volume in the series of works treating in detail of the origin of species ; while the part which treats of man, is an application of those principles to the human race which had hitherto only been discussed as regards other animals and plants, and would have formed a fitting companion volume to the *Origin of Species*. This rearrangement could easily be effected in a future edition, and would have many advantages ; and should a similar suggestion come from other quarters we hope Mr. Darwin will adopt it.

In entering upon a criticism of some portions of these volumes, I am compelled to touch upon certain topics on which I hold, and have published, views differing considerably from those maintained by Mr. Darwin ; and I am glad to have this opportunity of showing to what extent a study of his facts and arguments have modified my opinions. Before plunging into the intricate subject of " sexual selection," I must, however, make a few remarks on Mr. Darwin's use of the same term " instinct " for what seem to me very distinct things. He classes as instincts, hunger, self-preservation, the mother's love of her offspring, and the infant's power of sucking. The first is a sensation, the second acquired habit, the third an emotion, the fourth a pleasurable exercise of certain muscles—none of them instinct in the same sense as the cause of the migration of birds, of the building of platforms by apes, of the avoidance of poisonous fruits or the dread of snakes—all of which are specially mentioned as instincts. To go into the question of which of these latter are acquired habits or acquired knowledge, and which are truly instinctive, would lead us too far ; but it is certainly not in accordance with our author's usual precision of language on other topics, to use the same term for a simple sensation like hunger—for a faculty which may be experience or may be simple dislike acquired by natural selection, like the avoidance of poisonous fruits—and for all the mental processes involved in a highly complex operation like that of the construction of a bird's nest. It is no doubt mainly due to the poverty of our language that one word has been used for so many distinct things ; but as long as this is the case it is hardly possible to avoid confusion of ideas,about instinct.

In discussing the subject of sexual selection it would perhaps have been a more convenient, even if a less scientific arrangement, to have treated first of those groups in which the evidence is clearest and most decisive ; for Mr. Darwin is often obliged to refer to these in advance to strengthen his argument in the case of those inferior groups in which it is much more difficult to obtain evidence. I shall therefore first consider what is proved in the case of birds.

In birds sexual differences are both more generally the rule and more wonderfully varied in character than in any other class of animals. The males sometimes possess special weapons for fighting together ; more frequently they charm the female by vocal or instrumental music ; more frequently still they are ornamented with all sorts of crests, wattles, horns, air-sacs, plumes, and lengthened feathers springing from all parts of the body. They are extremely pugnacious ; they sing in rivalry, and they perform the most extraordinary antics and dances during the breeding season, exhibiting in the most curious and often unexpected manner all their peculiar adornments before the female. It is proved that in many cases they have a taste for colour and for novelty ; and some female domestic birds are shown to have had such a fondness for a peculiarly coloured male as to refuse to pair with any other. When in addition to this we consider that many birds are polygamous, and that in these the sexual differences are almost always greatest, we must admit that sexual selection would necessarily produce an effect in developing weapons, musical organs, or ornaments in one or both sexes. But while sexual selection has thus been doing its work, the still more powerful agency of natural selection has not been in abeyance, but has also modified one or both sexes in accordance with their conditions of life ; and these in the case of birds are somewhat different in the two sexes. Whole groups of birds are evidently coloured for protection, resembling the desert sands, or the green leaves, or the arctic snows, among which they live; and as we may be sure that variations tending to other colours have appeared in these birds, and as we have no reason to believe that in these groups only the females have been indifferent to such adornment, we must admit that natural selection has here checked the action of sexual selection. There are, however, an immense number of birds in which the female only is of dull brown or green tints, while the male is adorned with the most splendid colours; and there are also a very large number in which both sexes are equally or almost equally brilliant; and, with very rare exceptions, the rule is found to hold that the former class all build open nests, the latter all covered or hidden nests. The bright-coloured female birds are thus concealed while incubating, the dull-coloured are exposed. This very curious relation appeared to me to indicate that natural selection had been more powerful than the laws, whatever they are, which primarily determine the colours of birds; that the females had in one case been prevented from acquiring any considerable portion of the gay colouring of the males because it was hurtful to them, and in the other case had acquired it because, being concealed during incubation, it was no more hurtful to them than to the males. Mr. Darwin objects to this explanation of the facts. He maintains that the "laws of inheritance" determine whether colour or any other ornament appearing in one sex shall be transmitted to that sex only or to both. So far there is nothing to object to. But he goes further, and maintains that this tendency cannot be affected by natural selection, and that if a particular colour-variation begins to be transmitted to both sexes, the mode of transmission cannot, by natural selection, be changed, so that the colour may continue to be transmitted to the male to whom it is useful, but cease to be transmitted to the female to whom it is hurtful. Mr. Darwin admits that the law itself varies very frequently; for he gives numerous instances in which the different species of a genus exhibit all the possible modes of transmission, and as these have all descended from a common ancestor, the law has varied somewhat rapidly. He also says, " The equal transmission of characters to both sexes is the commonest form of inheritance," and we may therefore fairly assume that before diversity arose between the sexes it was the rule for both sexes to vary together. But he believes that, under these circumstances, it would be exceedingly difficult for natural selection to change the male alone, and he gives an imaginary illustration to exhibit this difficulty. He supposes a fancier to wish to make a breed of pigeons in which the males should be pale blue, the females remaining the usual slaty colour; and he says, " All that he could do would be to persevere in selecting every male pigeon which was in the least degree of a paler blue," and to match these with slaty females, the result being, of course, "either a mongrel piebald lot, or more probably the speedy and complete loss of the pale blue colour." But the supposed fancier has here gone quite the wrong way to work. His primary want is, not " blue males," but a breed in which there is a tendency to *differentiation of sex.* His proper plan, therefore, would be to look over as many sets as possible of the progeny of single pairs of pigeons till he found one in which a *differentiation of sex appeared in the right direction*, the males being lighter, the females darker, in however slight a degree. Breeding from these again, he would probably in a few generations find a greater differentiation occur, for we know that such changes in the mode of transmission have often occurred in nature; and only when he had obtained a breed in which the sexes were strongly differentiated, variations of colour occurring frequently in the male sex, rarely or not at all in the female, would it be advisable for him to begin selecting for the

exact tint of colour he desired in his males. Now, though nature may often do more in the way of selection than man, we can hardly believe that anything can be done by man's selection which may not be done as effectually by natural selection; and as it is admitted that the dull colours of the females sitting on open nests are a protection to them, and also that variations in the mode of transmission frequently occur, what is to prevent the females being modified in the way most advantageous to them for protection, while the males are being modified in the way most advantageous to them, by sexual selection? When the males of a species began to gain bright colours by sexual selection, and these colours were transmitted to the females till they became injurious, it may be fairly assumed that they would be transmitted in somewhat varying degrees, for Mr. Darwin states (p. 177, vol.ii.), that the degree of limitation differs in species of the same group; and as from mere association in the same locality individuals of the same family have a good chance of breeding together, the less brilliant females and more brilliant males of such families would often produce offspring in which the sexual differences were still greater, and these would have the best chance of surviving again to leave offspring. It is true that brilliant males of the same stock with brilliant females would have an equal chance of leaving descendants, but as the females of their families would be at a great disadvantage and would less frequently rear offspring, while the females of the differentiated families would be protected, the latter would soon be in a majority of two to one, and must inevitably supplant the former. This view enables us to understand many facts given by Mr. Darwin which seem difficulties on his own hypothesis. Thus the sexes of Culicidæ and Tabanidæ among flies, differ in the structure of the mouth in accordance with difference of habits; some male Cirrhipedes have lost almost all their external organs, while the female has retained hers; and female glow-worms, as well as many female moths, have lost their wings. Such varied adaptations of one sex alone could only occur if the rule were almost universal, that variations were limited to the sex in which they originally appeared; but we have seen that the contrary is nearer to the truth, and it seems more probable that the phenomenon of strictly limited sexual transmission was actually produced by natural selection as soon as the need arose for a differentiation of the sexes in organization, habits, or economy, than that it is an independent law. It evidently could have been so produced as well as the primary separation of the sexes which Mr. Darwin does not seem to doubt was effected by means of natural selection; and he appears to be unnecessarily depreciating the efficacy of his own first principle when he places limited sexual transmission beyond the range of its power.

Passing now to the lower animals—fishes, and especially insects—the evidence for sexual selection becomes comparatively very weak; and it seems doubtful if we are justified in applying the laws which prevail among the highly organized and emotional birds, to interpret somewhat analogous results in their case. The rivalry between males, either by fighting together or by emitting attractive sounds or odours, undoubtedly acts in the case of insects as well as in the higher animals; but it is quite different with the other form of sexual selection. This depends upon the appreciation of slight differences of colour by the female, and also by her having the power as well as the will to reject such males as are slightly inferior in attractions; and on both these points there is no direct evidence but what tells against Mr. Darwin's view. Thus, he informs us that " fresh females are often found paired with battered, faded, or dingy males," and breeders agree that in the case of the various silk-moths the female exerts no choice whatever, Dr. Wallace of Colchester stating that he frequently finds the most vigorous females of *Bombyx Cynthia* paired with stunted males. But the Bombyces are among the most elegantly coloured of all moths.

From the fact that many male butterflies may be seen pursuing or crowding round the same female, Mr. Darwin concludes that the females prefer one male to another, because, if this were not the case, the pairing must be left to mere chance, and this does not seem to him a probable event. But surely the male who finally obtains the female will be either the most vigorous, or the strongest-winged, or the most patient of the two or three suitors—the one who tires out or beats off the rest. The pairing therefore will not be left to chance, and it is probably by such struggles that the males of almost all butterflies have been rendered much stronger-winged than the females. Throughout the whole of the other orders of insects there is no direct evidence whatever of sexual selection as regards colour; for the colours are generally similar in both sexes, and the particular colours that occur seem to be often determined by the greater or less need of protection. Thus the stinging Hymenoptera are, as a rule, conspicuously coloured; as are large numbers of the Hemiptera, which are protected by their disgusting odour. Coleoptera are almost all palpably protected, either by resembling inanimate objects, by obscurity, by hard coats of mail, or by being distasteful to birds; and those of the two latter categories are almost all conspicuously coloured. It seems to me therefore, much more probable that the colours of insects are due to the same unknown laws which have produced the colours of caterpillars, than that they are due to sexual selection. In caterpillars we have almost all the classes of coloration found in perfect insects. We have protective and conspicuous tints; and among the latter we have spots, streaks, bands, and patterns, perfectly definite in character and of the most brilliantly contrasted hues. We have also many ornamental appendages; beautiful fleshy tubercles or tentacles, hard spines, beautifully coloured hairs arranged in tufts, brushes, starry clusters, or long pencils,—and horns on the head and tail, either single or double, pointed or clubbed. Now if all these beautiful and varied ornaments can be produced and rendered constant in each species, by some unknown cause quite independent of sexual selection, why cannot the same cause produce the colours and many of the ornaments of perfect insects, subjected as they are to so much greater variety of conditions than their larvæ? In the case of butterflies it is a curious fact that the females are often much more variable than the males. The females of *Papilio memnon* and *Diadema auge* are perhaps the most variable of all butterflies, consisting of scores of such different insects that they have over and over again been described as distinct species, while in both cases the males are very constant. Had the males been differentiated by sexual selection we should have expected them to be more variable, as they always are among insects as regards largely developed jaws, horns, or other weapons undoubtedly used for sexual purposes. In many groups of butterflies too, the males of the different species of a genus closely resemble each other, while the females differ considerably, so that it often happens that forms considered to be mere varieties as long as the males only are known, become recognised as good species when the females are discovered. This is the case generally in *Ornithoptera*, several groups of *Papilio, Adolias, Diadema;* and it is so exactly the reverse of what obtains in birds that we must hesitate to apply the same explanation to the two sets of phenomena.

There are two other difficulties in the way of accepting Mr. Darwin's wide generalization as to the agency of sexual selection in producing the greater part of the colour that adorns the animal world. How are we to believe that the action of an ever varying fancy for any slight change of colour could produce and fix the definite colours and markings which actually characterize species. Successive generations of female birds choosing any little variety of colour that occurred among their suitors would necessarily lead to a speckled or piebald and unstable result, not to the beautifully definite colours and markings we see. To the agency of natural selection there is no such bar. Each variation is unerringly selected or rejected according as it is useful or the reverse; and as conditions change but slowly, modifications will necessarily be carried on and accumulated till they reach their highest point of efficiency. But how can the individual tastes of hundreds of successive generations of female birds produce any such definite or constant effect? Some law of necessary development of colour in certain parts of the body and in certain hues is first required, and then perhaps, in the case of birds, the females might choose the successive improvements as they occurred; though, unless other variations were altogether prevented, it seems just as likely that they would mar the effect the law of development of colour was tending to produce.

The other objection is, that there are signs of such a tendency, which, taken in connection with the cases of caterpillars, of shells, and other very low organisms, may cover the whole ground in the case of insects, and render sexual selection of colour as unnecessary as it is unsupported by direct evidence. In many islands of the Malay Archipelago, species of widely different genera of butterflies differ, in precisely the same way as to colour or form, from allied species in other islands. The same thing occurs to a less degree in other parts of the world. Here we have indications of some local modifying influence which is certainly not sexual selection. So, the production in the males only of certain butterflies, of a peculiar neuration of the wings, of differently formed legs, and especially of groups of peculiarly formed scales only to be detected by microscopical examination, indicate the existence of some laws of development capable of differentiating the sexes other than sexual selection.

On the whole then it seems to me, that the kind of sexual selection which depends on the female preferring certain colours or ornaments in the male, has not been proved to exist in insects. Their colours are probably due to some as yet unknown causes; the differences of the sexes consisting, partly of a greater intensity of colouring in the male, due perhaps to his smaller size and greater vigour, and partly of more or less protective tints acquired by the female alone on account of her slower flight and greater need for protection while depositing her eggs. Many other points of great interest must be passed over, but sufficient has been said to enable the reader fairly to compare the facts and arguments previously adduced by myself with those now set forth by Mr. Darwin, and to form a judgment as to the comparative importance to be attached to sexual selection and the need of protection, in determining the sexual differences of colour in animals.

Having in the first part of this article made some objections to the theory of sexual selection in its application to man, I will now briefly notice Mr. Darwin's account of the probable mode in which man became developed from his brute ancestor. All the evidence goes to show, that the change from brute to man took place in some limited area, probably tropical. Here he lost his hairy covering, acquired his erect form and his wonderful brain, and became so far advanced in the arts and in morals that natural selection ceased to act upon his mere bodily organization. It is also probable that he learnt to speak language, discovered the use of fire, and perhaps even of canoes, before he spread over the earth, and before the several races of man were differentiated. The agency through which this vast transformation occurred was the struggle for existence and natural selection—a struggle first with other animals, and when they were surpassed between tribe and tribe; and this alone Mr. Darwin thinks would, under favourable conditions, raise man to his present high position in the organic scale.

In this view there are many difficulties. How is it possible to conceive, that during the enormous interval required to change a quadrumanous, hairy, speechless animal, into erect, smooth-skinned, large-brained, fire-using man, while the struggle for existence was most severe (for by the severity of the struggle alone he was raised), he yet never spread over the earth but remained concentrated in a limited area. Had he spread widely during the process of modification, divergence of character would inevitably have occurred, and we should have had several distinct species of man. Mr. Darwin argues that the fact of man, even at his lowest stage of civilisation and intellect, being able to maintain himself, surrounded by the most powerful and ferocious animals, is due to his large brain, which is thus of the most essential use to him. But almost all herbivorous animals also maintain themselves under similar conditions, with no special endowment of brains; and in South America the apparently helpless and almost idiotic sloth is not exterminated, though exposed to the attacks of pumas, tiger-cats, and harpy-eagles. Man could have acquired very little of his superiority to animals by a struggle with animals. "Natural selection does not produce absolute perfection but only relative perfection." We have to fall back therefore on the struggle with his fellows—family with family, tribe with tribe. But for this to be at all effectual, one of the most essential conditions is a large population inhabiting an extensive area, and this the

conditions of the problem deny to us. The vast amount of the superiority of man to his nearest allies is what is so difficult to account for. His absolute erectness of posture, the completeness of his nudity, the harmonious perfection of his hands, the almost infinite capacities of his brain, constitute a series of correlated advances too great to be accounted for by the struggle for existence of an isolated group of apes in a limited area. And Mr. Darwin himself gives hints of unknown causes which may have aided in the work. He says: "An unexplained residuum of change, perhaps a large one, must be left to the assumed uniform action of those unknown agencies which occasionally induce strongly marked and abrupt deviations of structure in our domestic productions." And again: "If these causes, whatever they may be, were to act more uniformly and energetically during a lengthened period (and no reason can be assigned why this should not sometimes occur), the result would probably be, not mere slight differences, but well marked, constant modifications."

In concluding this very imperfect account of one of the most remarkable works in the English language, it may be affirmed, that Mr. Darwin has all but demonstrated the origin of man by descent from some inferior animal form—that he has proved the vast importance of sexual influences in modifying the colours and the structure of the more highly organized animals—and that he has thrown fresh light upon the intricate question of the mode of development of the moral and intellectual nature of man. Yet it must be admitted that there are many difficulties in the detailed application of his views; and it seems probable that these can only be overcome by giving more weight to those unknown laws whose existence he admits, but to which he assigns an altogether subordinate part in determining the development of organic forms.

205

18

Reprinted from Q. Rev. London **131**:47–90 (1871)

The Descent of Man, and Selection in Relation to Sex.
By Charles Darwin, M.A., F.R.S., &c. 2 vols. London,
1871.

IN Mr. Darwin's last work we possess at length a complete and
thorough exposition of his matured views. He gives us
the results of the patient labour of many years' unremitting
investigation and of the application of a powerful and acute
intellect, combined with an extraordinarily active imagination,
to an unequalled collection of most varied, interesting and
important biological data. In his earlier writings a certain
reticence veiled, though it did not hide, his ultimate conclusions
as to the origin of our own species; but now all possibility of
misunderstanding or of a repetition of former disclaimers on the
part of any disciple is at an end, and the entire and naked truth
as to the logical consequences of Darwinism is displayed with a
frankness which we had a right to expect from the distinguished
author. What was but obscurely hinted in the 'Origin of
Species' is here fully and fairly stated in all its bearings and
without disguise. Mr. Darwin has, in fact, 'crowned the edifice,'
and the long looked for and anxiously awaited detailed statement of his views as to the human race is now unreservedly put
before us.

We rise from the careful perusal of this book with mingled
feelings of admiration and disappointment. The author's style
is clear and attractive—clearer than in his earlier works—and his
desire to avoid every kind of conscious misrepresentation is as
conspicuous as ever. The number of interesting facts brought
forward is as surprising as is the ingenuity often displayed in
his manipulation of them. Under these circumstances it is a
most painful task to have to point out grave defects and serious
shortcomings. Mr. Darwin, however, seems in his recent work
even more than in his earlier productions to challenge criticism,
and to have thrown out ideas and suggestions with a distinct
view to their subsequent modification by others. It is but an
act of fairness to call attention to this :—

'False facts,' says Mr. Darwin, 'are highly injurious to the progress
of science, for they often long endure; but false views, if supported
by some evidence, do little harm, as every one takes a salutary pleasure
in proving their falseness; and when this is done, one path towards
error is closed and the road to truth is often at the same time opened.'
—*Descent of Man*, vol. ii. p. 385.

Although we are unable to agree entirely with Mr. Darwin in
this remark, it none the less contains an undoubted truth. We
cannot

cannot agree, because we feel that a false theory which keenly solicits the imagination, put forward by a writer widely and deservedly esteemed, and which reposes on a multitude of facts difficult to verify, skilfully interwoven, and exceedingly hard to unravel, is likely to be very prejudicial to science. Nevertheless, science cannot make progress without the action of two distinct classes of thinkers: the first consisting of men of creative genius, who strike out brilliant hypotheses, and who may be spoken of as ' theorizers' in the good sense of the word ; the second, of men possessed of the critical faculty, and who test, mould into shape, perfect or destroy, the hypotheses thrown out by the former class.

Obviously important as it is that there should be such theorizers, it is also most important that criticism should clearly point out when a theory is really proved, when it is but probable, and when it is a mere arbitrary hypothesis. This is all the more necessary if, as may often and very easily happen, from being repeatedly spoken of, and being connected with celebrated and influential names, it is likely to be taken for very much more than it is really worth.

The necessity of caution in respect to this is clearly shown by Mr. Darwin's present work, in which ' sexual selection,' from being again and again referred to as if it had been proved to be a *vera causa*, may readily be accepted as such by the uninstructed or careless reader. For many persons, at first violently opposed through ignorance or prejudice to Mr. Darwin's views, are now, with scarcely less ignorance and prejudice, as strongly inclined in their favour.

Mr. Darwin's recent work, supplementing and completing, as it does, his earlier publications, offers a good opportunity for reviewing his whole position. We shall thus be better able to estimate the value of his convictions regarding the special subject of his present inquiry. We shall first call attention to his earlier statements, in order that we may see whether he has modified his views, and, if so, how far and with what results. If he has, even by his own showing and admission, been over-hasty and seriously mistaken previously, we must be the more careful how we commit ourselves to his guidance now. We shall endeavour to show that Mr. Darwin's convictions have undergone grave modifications, and that the opinions adopted by him now are quite distinct from, and even subversive of, the views he originally put forth. The assignment of the law of ' natural selection' to a subordinate position is virtually an abandonment of the Darwinian theory; for the one distinguishing feature of that theory was the all-sufficiency of ' natural selection.' Not
the

the less, however, ought we to feel grateful to Mr. Darwin for bringing forward that theory, and for forcing on men's minds, by his learning, acuteness, zeal, perseverance, firmness, and candour, a recognition of the probability, if not more, of evolution and of the certainty of the action of ' natural selection.' For though the ' survival of the fittest' is a truth which readily presents itself to any one who considers the subject, and though its converse, the destruction of the least fit, was recognised thousands of years ago, yet to Mr. Darwin, and (through Mr. Wallace's reticence) to Mr. Darwin alone, is due the credit of having first brought it prominently forward and demonstrated its truth in a volume which will doubtless form a landmark in the domain of zoological science.

We find even in the third edition of his ' Origin of Species' the following passages :—' Natural selection can act only by taking advantage of slight successive variations ; she can never take a leap, but must advance by short and slow steps ' (p. 214). Again he says :—' If it could be demonstrated that any complex organ existed, which could not possibly have been formed by numerous, successive, slight modifications, my theory would absolutely break down. But I can find out no such case ' (p. 208). He adds :—

' Every detail of structure in every living creature (making some little allowance for the direct action of physical conditions) may be viewed, either as having been of special use to some ancestral form, or as being now of special use to the descendants of this form—either directly, or indirectly through the complex laws of growth ;' and ' if it could be proved that any part of the structure of any one species had been formed for the exclusive good of another species, it would annihilate my theory, for such could not have been produced through natural selection' (p. 220).

It is almost impossible for Mr. Darwin to have used words by which more thoroughly to stake the whole of his theory on the non-existence or non-action of causes of any moment other than natural selection. For why should such a phenomenon ' annihilate his theory'? Because the very essence of his theory, as originally stated, is to recognise only the conservation of minute variations directly beneficial to the creature presenting them, by enabling it to obtain food, escape enemies, and propagate its kind. But once more he says :—

' We have seen that species at any one period are not indefinitely variable, and are not linked together by a multitude of intermediate gradations, partly because the process of natural selection will always be very slow, and will act, at any one time, only on a very few forms ; and partly because the very process of natural selection almost implies

implies the continual supplanting and extinction of preceding and intermediate gradations.'—P. 223.

Such are Mr. Darwin's earlier statements. At present we read as follows :—

' I now admit, after reading the essay by Nägeli on plants, and the remarks by various authors with respect to animals, more especially those recently made by Professor Broca, that in the earlier editions of my "Origin of Species" I probably attributed too much to the action of natural selection or the survival of the fittest.' . . . . ' I had not formerly sufficiently considered the existence of many structures which appear to be, as far as we can judge, neither beneficial nor injurious; and this I believe to be one of the greatest oversights as yet detected in my work.'—(' Descent of Man,' vol. i. p. 152.)

A still more remarkable admission is that in which he says, after referring to the action of both natural and sexual selection :—

' An unexplained residuum of change, perhaps a large one, must be left to the assumed action of those *unknown agencies*, which occasionally induce strongly marked and abrupt deviations of structure in our domestic productions.'—vol. i. p. 154.

But perhaps the most glaring contradiction is presented by the following passage :—

' No doubt man, as well as every other animal, presents structures, which as far as we can judge with our little knowledge, are not now of any service to him, nor have been so during any former period of his existence, either in relation to his general conditions of life, or of one sex to the other. Such structures cannot be accounted for by any form of selection, or by the inherited effects of the use and disuse of parts. We know, however, that many strange and strongly marked peculiarities of structure occasionally appear in our domesticated productions; and if the unknown causes which produce them were to act more uniformly, they would probably become common to all the individuals of the species.'—vol. ii. p. 387.

Mr. Darwin, indeed, seems now to admit the existence of internal, innate powers, for he goes on to say :—

' We may hope hereafter to understand something about the causes of such occasional modifications, especially through the study of monstrosities.' . . . . ' In the greater number of cases we can only say that the cause of each slight variation and of each monstrosity lies much more *in the nature or constitution of the organism** than in the nature of the surrounding conditions; though new and changed conditions certainly play an important part in exciting organic changes of all kinds.'

---

* The italics in the quotations from Mr. Darwin's book in this article are, in almost all cases, our's, and not the author's.

Also

Also, in a note (vol. i. p. 223), he speaks of 'incidental results of certain unknown differences in the constitution of the reproductive system.'

Thus, then, it is admitted by our author that we may have 'abrupt, strongly marked' changes, 'neither beneficial nor injurious' to the creatures possessing them, produced 'by unknown agencies' lying deep in 'the nature or constitution of the organism,' and which, if acting uniformly, would 'probably' modify similarly 'all the individuals of a species.' If this is not an abandonment of 'natural selection,' it would be difficult to select terms more calculated to express it. But Mr. Darwin's admissions of error do not stop here. In the fifth edition of his 'Origin of Species' (p. 104) he says, 'Until reading an able and valuable article in the "North British Review" (1867), I did not appreciate how rarely single variations, whether slight or strongly marked, could be perpetuated.' Again: he was formerly 'inclined to lay much stress on the principle of protection, as accounting for the less bright colours of female birds' ('Descent of Man,' vol. ii. p. 198) ; but now he speaks as if the correctness of his old conception of such colours being due to protection was unlikely. 'Is it probable,' he asks, 'that the head of the female chaffinch, the crimson on the breast of the female bullfinch, —the green of the female chaffinch,—the crest of the female golden-crested wren, have all been rendered less bright by the slow process of selection for the sake of protection? *I cannot think so*' (vol. ii. p. 176.)

Once more Mr. Darwin shows us (vol. i. p. 125) how he has been over-hasty in attributing the development of certain structures to reversion. He remarks, 'In my "Variations of Animals under Domestication" (vol. ii. p. 57) I attributed the not very rare cases of supernumerary mammæ in women to reversion.' 'But Professor Preyer states that *mammæ erraticæ* have been known to occur in other situations, even on the back; so that the force of my argument is greatly weakened or perhaps quite destroyed.'

Finally, we have a postscript at the beginning of the second volume of the 'Descent of Man' which contains an avowal more remarkable than even the passages already cited. He therein declares :—

'I have fallen into a serious and unfortunate error, in relation to the sexual differences of animals, in attempting to explain what seemed to me a singular coincidence in the late period of life at which the necessary variations have arisen in many cases, and the late period at which sexual selection acts. The explanation given is wholly erroneous,

erroneous, as I have discovered by working out an illustration in figures.'

While willingly paying a just tribute of esteem to the candour which dictated these several admissions, it would be idle to dissemble, and disingenuous not to declare, the amount of distrust with which such repeated over-hasty conclusions and erroneous calculations inspire us. When their Author comes before us anew, as he now does, with opinions and conclusions still more startling, and calculated in a yet greater degree to disturb convictions reposing upon the general consent of the majority of cultivated minds, we may well pause before we trust ourselves unreservedly to a guidance which thus again and again declares its own reiterated fallibility. Mr. Darwin's conclusions may be correct, but we feel we have now indeed a right to demand that they shall be proved before we assent to them ; and that since what Mr. Darwin before declared '*must* be,' he now admits not only to be unnecessary but untrue, we may justly regard with extreme distrust the numerous statements and calculations which, in the ' Descent of Man,' are avowedly recommended by a mere '*may* be.' This is the more necessary, as the Author, starting at first with an avowed hypothesis, constantly asserts it as an undoubted fact, and claims for it, somewhat in the spirit of a theologian, that it should be received as an article of faith. Thus the formidable objection to Mr. Darwin's theory, that the great break in the organic chain between man and his nearest allies, which cannot be bridged over by any extinct or living species, is answered simply by an appeal ' to a *belief* in the general principle of evolution' (vol. i. p. 200), or by a confident statement that ' we have *every reason to believe* that breaks in the series are simply the result of many forms having become extinct' (vol. i. p. 187). So, in like manner, we are assured that ' the early progenitors of man were, *no doubt*, once covered with hair, both sexes having beards ; their ears were pointed and capable of movement; and their bodies were provided with a tail, having the proper muscles ' (vol. i. p. 206). And, finally, we are told, with a dogmatism little worthy of a philosopher, that, ' *unless we wilfully close our eyes*,' we must recognise our parentage (vol. i. p. 213).

These are hard words ; and, even at the risk of being accused of wilful blindness, we shall now proceed, with an unbiassed and unprejudiced mind, to examine carefully the arguments upon which Mr. Darwin's theory rests. Must we acknowledge that ' man with all his noble qualities, with sympathy which feels for the most debased, with benevolence which extends not only to
other

other men but to the humblest living creature, with his god-like intellect which has penetrated into the movements and constitution of the solar system,' must we acknowledge that man ' with all these exalted powers' is descended from an Ascidian? Is this a scientific truth resting on scientific evidence, or is it to be classed with the speculations of a bygone age?

With regard to the Origin of Man, Mr. Darwin considers that both ' natural selection' and ' sexual selection' have acted. We need not on the present occasion discuss the action of natural selection; but it will be necessary to consider that of ' sexual selection' at some length. It plays a very important part in the ' descent of man,' according to Mr. Darwin's views. He maintains that we owe to it our power of song and our hairlessness of body, and that also to it is due the formation and conservation of the various races and varieties of the human species. In this matter then we fear we shall have to make some demand upon our readers' patience. ' Sexual selection' is the corner-stone of Mr. Darwin's theory. It occupies three-fourths of his two volumes ; and unless he has clearly established this point, the whole fabric falls to the ground. It is impossible, therefore, to review the book without entering fully into the subject, even at the risk of touching upon some points which, for obvious reasons, we should have preferred to pass over in silence.

Under the head of 'sexual selection' Mr. Darwin includes two very distinct processes. One of these consists in the action of superior strength or activity, by which one male succeeds in obtaining possession of mates and in keeping away rivals. This is, undoubtedly, a *vera causa;* but may be more conveniently reckoned as one kind of ' natural selection' than as a branch of ' sexual selection.' The second process consists in alleged preference or choice, exercised freely by the female in favour of particular males on account of some attractiveness or beauty of form, colour, odour, or voice, which such males may possess. It is this second kind of ' sexual selection' (and which alone deserves the name) that is important for the establishment of Mr. Darwin's views, but its valid action has to be proved.

Now, to prove the existence of such a power of choice Mr. Darwin brings forward a multitude of details respecting the sexual phenomena of animals of various classes ; but it is the class of birds which is mainly relied on to afford evidence in support of the exercise of this power of choice by female animals. We contend, however, that not only is the evidence defective even here, but that much of his own evidence is in direct opposition to his views. While the unquestionable fact, that male sexual characters (horns, mane, wattles,

wattles, &c., &c.) have been developed in many cases where sexual selection has certainly not acted, renders it probable, *a priori*, that the unknown cause which has operated in these numerous cases has operated in those instances also which seem to favour the hypothesis supported by Mr. Darwin. Still he contends that the greater part of the beauty and melody of the organic world is due exclusively to this selective process, by which, through countless generations, the tail of the peacock, the throat of the humming-bird, the song of the nightingale, and the chirp of the grasshopper have been developed by females, age after age, selecting for their mates males possessing in a more and more perfect degree characters which must thus have been continually and constantly preferred.

Yet, after all, Mr. Darwin concedes *in principle* the very point in dispute, and yields all for which his opponents need argue, when he allows that beautiful and harmonious variations may occur *spontaneously* and *at once*, as in the dark or spangled bars on the feathers of Hamburgh fowls (' Descent of Man,' vol. i. p. 281). For what difference is there, other than mere difference of degree, between the spontaneous appearance of a few beautiful new feathers with harmonious markings and the spontaneous appearance of a whole beautiful clothing like that of the Tragopans ?

Again, on Mr. Darwin's own showing, it is manifest that male sexual characters, such as he would fain attribute to sexual selection, may arise without any such action whatever. Thus he tells us, ' There are breeds of the sheep and goat, in which the horns of the male differ greatly in shape from those of the female ;' and ' with tortoise-shell cats, the females alone, as a general rule, are thus coloured, the males being rusty-red ' (vol. i. p. 283). Now, if these cats were only known in a wild state, Mr. Darwin would certainly bring them forward amongst his other instances of alleged sexual selection, though we now know the phenomenon is not due to any such cause. A more striking instance, however, is the following :—' With the pigeon, the sexes of the parent species do not differ in any external character; nevertheless, in certain domesticated breeds the male is differently coloured from the female. The wattle in the English carrier-pigeon and the crop in the pouter are more highly developed in the male than in the female ;' and ' this has arisen, not from, but rather *in opposition to*, the wishes of the breeder ;' which amounts to a positive demonstration that sexual characters may arise spontaneously, and, be it noted, in the class of birds.

The uncertainty which besets these speculations of Mr. Darwin is evident at every turn. What at first could be thought a
better

better instance of sexual selection than the light of the glowworm, exhibited to attract her mate? Yet the discovery of luminous larvæ, which of course have no sexual action, leads Mr. Darwin to observe: 'It is very doubtful whether the primary use of the light is to guide the male to the female' (vol. i. p. 345). Again, as to certain British field-bugs, he says: 'If in any species the males had differed from the females in an analogous manner, we *might have been justified* in attributing such conspicuous colours to sexual selection with transference to both sexes' (vol. i. p. 350). As to the stridulating noises of insects (which is assumed to be the result of sexual selection), Mr. Darwin remarks of certain Neuroptera:—' It is rather surprising that both sexes should have the power of stridulating, as the male is winged and the female wingless' (vol. i. p. 366); and he is again surprised to find that this power is not a sexual character in many Coleoptera (vol. i. p. 382).

Moths and butterflies, however, are the insects which Mr. Darwin treats of at the greatest length in support of sexual selection. Yet even here he supplies us with positive evidence that in certain cases beauty does not charm the female. He tells us:—

' Some facts, however, are opposed to the belief that female butterflies prefer the more beautiful males; thus, as I have been assured by several observers, fresh females may frequently be seen paired with battered, faded, or dingy males.'—vol. i. p. 400.

As to the Bombycidæ he adds:—

' The females lie in an almost torpid state, and appear not to evince the least choice in regard to their partners. This is the case with the common silk-moth (*B. mori*). Dr. Wallace, who has had such immense experience in breeding *Bombyx cynthia*, is convinced that the females evince no choice or preference. He has kept above 300 of these moths living together, and has often found the most vigorous females mated with stunted males.'

Nevertheless, we do not find, for all this, any defect of colour or markings, for, as Mr. Alfred Wallace observes (*Nature*, March 15th, 1871, p. 182), 'the Bombyces are amongst the most elegantly coloured of all moths.'

Mr. Darwin gives a number of instances of sexual characters, such as horns, spines, &c., in beetles and other insects; but there is no fragment of evidence that such structures are in any way due to feminine caprice. Other structures are described and figured which doubtless do aid the sexual act, as the claws of certain Crustacea; but these are often of such size and strength (*e. g.* in *Callianassa* and *Orchestia*) as to render any power of
choice

choice on the part of the female in the highest degree incredible.

Similarly with the higher classes, *i.e.* Fishes, Reptiles, and Beasts, we have descriptions and representations of a number of sexual peculiarities, but no evidence whatever that such characters are due to female selection. Often we have statements which conflict strongly with a belief in any such action. Thus, *e. g.*, Mr. Darwin quotes Mr. R. Buist, Superintendent of Fisheries, as saying that male salmon

' Are constantly fighting and tearing each other on the spawning-beds, and many so injure each other as to cause the death of numbers, many being seen swimming near the banks of the river in a state of exhaustion, and apparently in a dying state.' . . . 'The keeper of Stormontfield found in the northern Tyne about 300 dead salmon, all of which with one exception were males; and he was convinced that they had lost their lives by fighting.'—vol. ii. p. 3.

The female's choice must here be much limited, and the only kind of sexual selection which can operate is that first kind, determined by combat, which, we before observed, must rather be ranked as a kind of 'natural selection.' Even with regard to this, however, we may well hesitate, when Mr. Darwin tells us, as he does, that seeing the habitual contests of the males, ' it is surprising that they have not generally become, through the effects of sexual selection, larger and stronger than the females;' and this the more as ' the males suffer from their small size,' being 'liable to be devoured by the females of their own species' (vol. ii. p. 7). The cases cited by our Author with regard to fishes, do not even tend to prove the existence of sexual selection, and the same may be said as to the numerous details given by him about Reptiles and Amphibians. Nay, rather the facts are hostile to his views. Thus, he says himself, 'It is surprising that frogs and toads should not have acquired more strongly-marked sexual differences; for though cold-blooded, their passions are strong' (vol. ii. p. 26). But he cites a fact, than which it would be difficult to find one less favourable to his cause. He adds: 'Dr. Günther informs me that he has several times found an unfortunate female toad dead and smothered from having been so closely embraced by three or four males.' If female selection was difficult in the case of the female salmon, it must be admitted to have been singularly infelicitous to the female toad.

We will now notice some facts brought forward by Mr. Darwin with regard to beasts. And first, as to the existence of choice on the part of the females, it may be noted that ' Mr. Bienkiron, the greatest breeder of race-horses in the world, says that

that stallions are so frequently capricious in their choice, rejecting one mare and without any apparent cause taking to another, that various artifices have to be habitually used.' 'He has *never known a mare to reject a horse;*' though this has occurred in Mr. Wright's stable.

Some of the most marked sexual characters found amongst mammals, are those which exist in apes. These are abundantly noticed by Mr. Darwin, but his treatment of them seems to show his inability to bring them within the scope of his theory.

It is well known that certain apes are distinguished by the lively colours or peculiarities as to hair possessed by the males, while it is also notorious that their vastly superior strength of body and length of fang, would render resistance on the part of the female difficult and perilous, even were we to adopt the utterly gratuitous supposition, that at seasons of sexual excitement the female shows any disposition to coyness. Mr. Darwin has no facts to bring forward to prove the exercise of any choice on the part of female apes, but gives in support of his views the following remarkable passage :—

'Must we attribute to mere purposeless variability in the male all these appendages of hair and skin ? It cannot be denied that this is possible; for, with many domesticated quadrupeds, certain characters, apparently not derived through reversion from any wild parent-form, have appeared in, and are confined to, the males, or are more largely developed in them than in the females,—for instance, the hump in the male zebu-cattle of India, the tail in fat-tailed rams, the arched outline of the forehead in the males of several breeds of sheep, the mane in the ram of an African breed, and, lastly, the mane, long hairs on the hinder legs, and the dewlap in the male alone of the Berbura goat.'— vol. ii. p. 284.

If these are due, as is probable, to simple variability, then, he adds,—

' It would appear reasonable to extend the same view to the many analogous characters occurring in animals under a state of nature. Nevertheless I cannot persuade myself that this view is applicable in many cases, as in that of the extraordinary development of hair on the throat and fore-legs of the male Ammotragus, or of the immense beard of the Pithecia (monkey).'—vol. ii. p. 285.

But one naturally asks, Why not? Mr. Darwin gives no reason (if such it may be called) beyond that implied in the gratuitous use of the epithet ' purposeless ' in the passage cited, and to which we shall return.

In the Rhesus monkey the female appears to be more vividly coloured than the male ; therefore Mr. Darwin infers (grounding his

his inference on alleged phenomena in birds) that sexual selection is *reversed*, and that in this case the male selects. This hypothetical reversion of a hypothetical process to meet an exceptional case will appear to many rash indeed, when they reflect that as to teeth, whiskers, general size, and superciliary ridges this monkey 'follows the common rule of the male excelling the female' (vol. ii. p. 294).

To turn now to the class on which Mr. Darwin especially relies, we shall find that even Birds supply us with numerous instances which conflict with his hypothesis. Thus, speaking of the battling of male waders, our author tells us:—'Two were seen to be thus engaged for half an hour, until one got hold of the head of the other, which would have been killed had not the observer interfered; the female all the time looking on as a quiet spectator' (vol. ii. p. 41). As these battles must take place generally in the absence of spectators, their doubtless frequently fatal termination must limit greatly the power of selection Mr. Darwin attributes to the females. The same limit is certainly imposed in the majority of Gallinaceous birds, the cocks of which fight violently; and there can be little doubt but that, as an almost invariable rule, the victorious birds mate with the comparatively passive hens.

Again, how can we explain, on Mr. Darwin's hypothesis, the existence of distinguishing male sexual marks, where it is the male and not the female bird which selects? Yet the wild turkey-cock, a distinguished bird enough, is said by Mr. Darwin (vol. ii. p. 207) to be courted by the females; and he quotes (vol. ii. p. 120) Sir R. Heron as saying, 'that with peafowl, the first advances are always made by the female.' And of the capercailzie he says, 'the females flit round the male while he is parading, and solicit his attention.'

But though, of course, the sexual instinct always seeks its gratification, does the female *ever* select a particular plumage? The strongest instance given by Mr. Darwin is as follows:—

'Sir R. Heron during many years kept an account of the habits of the peafowl, which he bred in large numbers. He states that the hens have frequently great preference for a particular peacock. They were all so fond of an old pied cock, that one year, when he was confined though still in view, they were constantly assembled close to the trellice-walls of his prison, and would not suffer a japanned peacock to touch them. On his being let out in the autumn, the oldest of the hens instantly courted him, and was successful in her courtship. The next year he was shut up in a stable, and then the hens all courted his rival. This rival was a japanned or black-winged peacock, which to our eyes is a more beautiful bird than the common kind.'—vol. ii. p. 119.

Now

Now no one disputes as to birds showing preferences one for another, but it is quite a gratuitous suggestion that the pied plumage of the venerable paterfamilias was *the* charm which attracted the opposite sex; and even if such were the case, it would seem (from Mr. Darwin's concluding remark) to prove either that the peahen's taste is so different from ours, that the peacock's plumage could never have been developed by it, or (if the taste of these peahens was different from that of most peahens) that such is the instability of a vicious feminine caprice, that no constancy of coloration could be produced by its selective action.

Mr. Darwin bases his theory of sexual selection greatly on the fact that the male birds display the beauty of their plumage with elaborate parade and many curious and uncouth gestures. But this display is not exclusively used in attracting and stimulating the hens. Thus he admits that 'the males will sometimes display their ornaments when *not* in the presence of the females, as occasionally occurs with the grouse at their balz-places, and as may be noticed with the peacock; this latter bird, however, evidently wishes for a spectator *of some kind*, and will show off his finery, as I have often seen, before poultry or even pigs' (vol. ii. p. 86). Again, as to the brilliant *Rupicola crocea*, Sir R. Schomburgk says: 'A *male* was capering to the apparent delight of several *others*' (vol. ii. p. 87).

From the fact of 'display' Mr. Darwin concludes that 'it is obviously probable that the females appreciate the beauty of their suitors' (vol. ii. p. 111). Our Author, however, only ventures to call it 'probable,' and he significantly adds: 'It is, however, difficult to obtain direct evidence of their capacity to appreciate beauty.' And again he says of the hen bird: 'It is not probable that she consciously deliberates; but she is most excited or attracted by the most beautiful, or melodious, or gallant males' (vol. ii. p. 123). No doubt the plumage, song, &c., all play their parts in aiding the various processes of life; but to stimulate the sexual instinct, even supposing this to be the object, is one thing —to supply the occasion for the exercise of a power of choice is quite another. Certainly we cannot admit what Mr. Darwin affirms (vol. ii. p. 124), that an 'even occasional preference by the female of the more attractive males would almost certainly lead to their modification.'

A singular instance is given by Mr. Darwin (vol. ii. p. 111) in support of his view, on the authority of Mr. J. Weir. It is that of a bullfinch which constantly attacked a reed-bunting, newly put into the aviary; and this attack is attributed to a sort of jealousy on the part of the blackheaded bullfinch of the black
head

head of the bunting.   But the bullfinch could hardly be aware of the colour of the top of its own head!

Mr. Wallace accounts for the brilliant colours of cater-pillars and many birds in another way.   The caterpillars which are distasteful must have gained if 'some outward sign indicated to their would - be destroyer that its prey was a disgusting morsel.'   As to birds, he believes that brilliance of plumage is developed where not hurtful, and that the generally more sober plumage of the hens has been produced by natural selection, killing off the more brilliant ones exposed during incubation to trying conditions.

Now as Mr. Wallace disposes of Mr. Darwin's views by his objections, so Mr. Darwin's remarks tend to refute Mr. Wallace's positions, and the result seems to point to the existence of some unknown innate and internal law which determines at the same time both coloration and its transmission to either or to both sexes. At the same time these authors, indeed, show the *harmony* of natural laws and processes one with another, and their mutual interaction and aid.

It cannot be pretended that there is any evidence for sexual selection except in the class of Birds.   Certain of the pheno-mena which Mr. Darwin generally attributes to such selection must be due, in some other classes, to other causes, and there is no *proof* that sexual selection acts, *even* amongst birds.

But in other classes, as we have seen, sexual characters are as marked as they are in the feathered group.   Mr. Darwin, indeed, argues that birds select, and assumes that their sexual charac-ters have been produced by such sexual selection, and that, therefore, the sexual characters of beasts have been similarly evolved.   But we may turn the argument round and say that sexual characters not less strongly marked exist in many beasts, reptiles, and insects, which characters cannot be due to sexual selection; that it is, therefore, probable the sexual characters of birds are not due to sexual selection either, but that some unknown internal cause has equally operated in each case.   The matter, indeed, stands thus: Of animals possessing sexual characters there are some in which sexual selection cannot have acted; others in which it may possibly have acted; others again in which, according to Mr. Darwin, it has certainly acted.   It is a somewhat singular conclusion to deduce from this that sexual selection is the one universal cause of sexual characters, when similar effects to those which it is supposed to cause take place in its absence.

But, indeed, what are the data on which Mr. Darwin relies as regards birds?   As before said, they are 'display' by the
males,

males, the 'greater brilliancy and ornamentation of these,' and the 'occasional preference' by females in confinement for particular males. Is there here any sufficient foundation for such a superstructure? In the first place, in insects, *e. g.* butterflies, we have often many brilliant males crowding in pursuit of a single female. Yet, as Mr. Wallace justly observes, 'Surely the male who finally obtains the female will be either the most vigorous, or the strongest-winged, or the most patient—the one who tires out or beats off the rest.' Similarly in birds strength and perseverance will, no doubt, generally reward the suitor possessing those qualities. Doubtless, also, this will generally be the most beautiful or most melodious; but this will simply be because extra beauty of plumage, or of song, will accompany supereminent vigour of constitution and fulness of vitality. What has been before said as to the fierce combats of cock-birds must be borne in mind.

But that internal spontaneous powers *are* sufficient to produce all the most varied or bizarre sexual characters which any birds exhibit, is actually demonstrated by the class of insects, especially caterpillars which from their sexless undeveloped state can have nothing to do with the kind of selection Mr. Darwin advocates. Yet amongst caterpillars we not only find some ornamented with spots, bands, stripes, and curious patterns, 'perfectly definite in character and of the most brilliantly contrasted hues. We have also many ornamental appendages; beautiful fleshy tubercles or tentacles, hard spines, beautifully coloured hairs arranged in tufts, brushes, starry clusters, or long pencils, and horns on the head and tail, either single or double, pointed or clubbed.' Mr. Wallace adds, 'Now if all these beautiful and varied ornaments can be produced and rendered constant in each species by some unknown cause quite independent of sexual selection, why cannot the same cause produce the colours and many of the ornaments of perfect insects;'—we may also add, the colours and ornaments of all other animals, including birds?

There is, however, another reason which induces Mr. Darwin to accept sexual selection; and it is probably this which, in his mind, mainly gives importance to the facts mentioned as to the plumage and motions of birds. He says of 'display,' 'It is incredible that all this display should be purposeless' (vol. ii. p. 399); and again (vol. ii. p. 93), he declares that any one who denies that the female Argus pheasant can appreciate the refined beauty of the plumage of her mate, 'will be compelled to admit that the extraordinary attitudes assumed by the male during the act of courtship, by which the wonderful beauty of his plumage

is

is fully displayed, are purposeless; and this is a conclusion which I for one will never admit.' It seems then that it is this imaginary necessity of attributing purposelessness to acts, which determines Mr. Darwin to attribute that peculiar and special purpose to birds' actions which he does attribute to them. But surely this difficulty is a mere chimæra. Let it be granted that the female does not select; yet the display of the male may be useful in supplying the necessary degree of stimulation to her nervous system, and to that of the male. Pleasurable sensation, perhaps very keen in intensity, may thence result to both. There would be no difficulty in suggesting yet other purposes if we were to ascend into higher speculative regions. Mr. Darwin has given us in one place a very remarkable passage; he says:—

'With respect to female birds feeling a preference for particular males, we must bear in mind that we can judge of choice being exerted, only by placing ourselves in imagination in the same position. If an inhabitant of another planet were to behold a number of young rustics at a fair, courting and quarrelling over a pretty girl, like birds at one of their places of assemblage, he would be able to infer that she had the power of choice only by observing the eagerness of the wooers to please her, and to display their finery.'—vol. ii. p. 122.

Now here it must be observed that, as is often the case, Mr. Darwin assumes the very point in dispute, unless he means by 'power of choice' mere freedom of physical power. If he means an internal, mental faculty of choice, then the observer could attribute such power to the girl only if he had reason to attribute to the rustics an intellectual and moral nature similar in kind to that which he possessed himself. Such a similarity of nature Mr. Darwin, of course, does attribute to rational beings and to brutes; but those who do not agree with him in this would require other tests than the presence of ornaments, and the performance of antics and gestures unaccompanied by any evidence of the faculty of articulate speech.

Such, then, is the nature of the evidence on which sexual selection is supposed to rest. To us the action of sexual selection scarcely seems more than a possibility, the evidence rarely raising it to probability. It cannot be a 'sufficient cause' to account for the phenomena which it is intended to explain, nor can it even claim to be taken as a *vera causa* at all. Yet Mr. Darwin again and again speaks as if its reality and cogency were indisputable.

As to the alleged action of natural selection on our own species we may mention two points.

First, as to the absence of hair. This is a character which Mr. Darwin admits cannot be accounted for by 'natural selection,'

tion,' because manifestly not beneficial ; it is therefore attributed to 'sexual selection,' incipient man being supposed to have chosen mates with less and less hairy bodies ; and the possibility of such action is thought by Mr. Darwin to be supported by the fact that certain monkeys have parts of the body naked. Yet it is a fact that the higher apes have not this nakedness, or have it in a much smaller degree.

Secondly, as to the races of mankind, Mr. Darwin's theory, indeed, requires the alternation of constancy and caprice to account for the selection and the conservation of marked varieties. In order that each race may possess and preserve its own ideal standard of beauty we require the truth of the hypothesis that ' certain tastes may in the course of time become inherited ;' and yet Mr. Darwin candidly admits (vol. ii. p. 353) : 'I know of no evidence in favour of this belief.' On the other hand, he says (p. 370), As soon as tribes exposed to different conditions came to vary, ' each isolated tribe would form for itself a slightly different standard of beauty,' which ' would gradually and inevitably be increased to a greater and greater degree.' But why have not the numerous tribes of North American Indians diverged from each other more conspicuously, inhabiting, as they do, such different climates, and surrounded by such diverse conditions?

Again, far from each race being bound in the trammels of its own features, all cultivated Europeans, whether Celts, Teutons, or Slaves, agree in admiring the Hellenic ideal as the highest type of human beauty.

We may now pass on to the peculiarities of man's bodily frame, and the value and signification of the resemblances presented by it to the various structures which are found to exist in lower members of the animal kingdom.

Mr. Darwin treats us to a very interesting account, not only of man's anatomy, but also of the habits, diseases, and parasites (internal and external) of man, together with the process of his development. He points out (vol. i. p. 11) not only the close similarity even of cerebral structure between man and apes, but also how the same animals are ' liable to many of the same non-contagious diseases as we are ; thus Rengger, who carefully observed for a long time the *Cebus Azaræ* in its native land, found it liable to catarrh, with the usual symptoms, and which when often recurrent, led to consumption. These monkeys suffered also from apoplexy, inflammation of the bowels, and cataract in the eye. The younger ones, when shedding their milk-teeth, often died from fever. Medicines produced the same effect on them as on us. Many kinds

kinds of monkeys have a strong taste for tea, coffee, and spirituous liquors; they will also, as I have myself seen, smoke tobacco with pleasure.' He also tells us of baboons which, after taking too much beer, 'on the following morning were very cross and dismal, held their aching heads with both hands, and wore a most pitiable expression: when beer or wine was offered them, they turned away with disgust, but relished the juice of lemons.' He notices, besides, the process of development in man with the transitory resemblances it exhibits to the immature conditions of other animals, and he mentions certain muscular abnormalities.

Mr. Darwin also brings forward an observation of Mr. Woolner, the sculptor, as to a small projection of the helix or outermost fold of the human ear, which projection 'we may safely conclude' to be 'a vestige of formerly pointed ears—which occasionally reappears in man' (vol. i. p. 23). Very many other interesting facts are noted which it would be superfluous here to recapitulate. It is, however, in connexion with man's bodily structure and its resemblances that we have observed slight errors on the part of Mr. Darwin, which it may be as well to point out; though it should be borne in mind that he does not profess to be in any sense an anatomist. Thus, at vol. i. p. 28, he mistakes the supra-condyloid foramen of the humerus for the inter-condyloid perforation. Did the former condition frequently occur in man—as, through this mistake, he asserts—it would be remarkable indeed, as it is only found in the lower monkeys and not in the higher. A more singular mistake is that of the malar bone for the premaxilla (vol. i. p. 124).

To return to the bodily and other characters enumerated at such length by Mr. Darwin. They may, and doubtless they will, produce a considerable effect on readers who are not anatomists, but in fact the whole and sole result is to show that man *is* an animal. That he is such is denied by no one, but has been taught and accepted since the time of Aristotle. We remember on one occasion meeting at a dinner-table a clever medical man of materialistic views. He strongly impressed the minds of some laymen present by an elaborate statement of the mental phenomena following upon different injuries, or diseased conditions of different parts of the brain, until one of the number remarked as a climax, 'Yes; and when the brain is entirely removed the mental phenomena cease altogether'—the previous observations having only brought out vividly what no one denied, viz., that during this life a certain integrity of bodily structure is requisite for the due exercise of the mental powers. Thus Mr. Darwin's remarks are merely an elaborate statement of
                                                                    what

what all admit, namely, that man is an animal. They further imply, however, that he is no more than an animal, and that the mode of origin of his visible being must be the mode of his origin as a whole—a conclusion of which we should not question the legitimacy if we could accept Mr. Darwin's views of man's mental powers.

All that can be said to be established by our author is, that if the various kinds of lower animals have been evolved one from the other by a process of natural generation or evolution, then it becomes highly probable *a priori* that man's body has been similarly evolved; but this, in such a case, becomes equally probable from the admitted fact that he is an animal at all.

The evidence for such a process of evolution of man's body amounts, however, only to an *a priori* probability, and might be reconciled with another mode of origin if there were sufficient reason (of another kind) to justify a belief in such other mode of origin. Mr. Darwin says:—'It is only our natural prejudice, and that arrogance which made our forefathers declare that they were descended from demi-gods, which leads us to demur to this conclusion' (vol. i. p. 32). But this is not the case; for many demur to his conclusion because they believe that to accept his view would be to contradict other truths which to them are far more evident.

He also makes the startling assertion that to take any other view than his as to man's origin, 'is to admit that our own structure and that of all the animals around us, is a mere snare laid to entrap our judgment' (vol. i. p. 32). Mr. Darwin is, we are quite sure, far enough from pretending that he has exhausted the possibilities of the case, and yet could anything but a conviction that the whole field had been explored exhaustively, justify such an assertion? If, without such a conviction, it were permissible so to dogmatize, every theorizer who had attained to a plausible explanation of a set of phenomena might equally make use of the assertion, and say, until a better explanation was found, that to doubt him would be to attribute duplicity to the Almighty.

In tracing man's origin Mr. Darwin is again betrayed into slight inaccuracies. Thus, in combating the position, advanced in this 'Review,'* that the hands of apes had been preformed (with a view to man) in a condition of perfection beyond their needs, he says :—

'On the contrary, I see no reason to doubt that a more perfectly constructed hand would have been an advantage to them, provided, and it is important to note this, that their hands had not thus been

---

* See 'Quarterly Review,' April, 1869, p. 392.

rendered

rendered less well adapted for climbing trees. We may suspect that a perfect hand would have been disadvantageous for climbing ; as the most arboreal monkeys in the world, namely Ateles in America and Hylobates in Asia, either have their thumbs much reduced in size and even rudimentary, or their fingers partially coherent, so that their hands are converted into grasping-hooks.'—vol. i. p. 140.

In a note, Mr. Darwin refers to the Syndactyle Gibbon as having two of the digits coherent. But these digits are not, as he supposes, digits of the hand but toes. Moreover, though doubtless the Gibbons and spider-monkeys are admirably organized for their needs, yet it is plain that a well-developed thumb is no impediment to climbing, for the strictly arboreal Lemurs are exceedingly well furnished in this respect. Again he says (vol. i. p. 143) of the Gibbons, that they, ' without having been taught, can walk or run upright with tolerable quickness, though they move awkwardly, and much less securely than man.' This is a little misleading, inasmuch as it is not stated that this upright progression is effected by placing the enormously long arms behind the head or holding them out backwards as a balance in progression.

We have already seen that Mr. Darwin tries to account for man's hairlessness by the help of ' sexual selection.' He also, however, speculates as to the possibility of his having lost it through heat of climate, saying :—' Elephants and rhinoceroses are almost hairless ; and as certain extinct species which formerly lived under an arctic climate were covered with long wool or hair, it would almost appear as if the existing species of both genera had lost their hairy covering from exposure to heat ' (vol. i. p. 148).

This affords us a good example of hasty and inconclusive speculation. Surely it would be as rational to suppose that the arctic species had *gained* their coats as that the tropical species had lost theirs. But over-hasty conclusions are, we regret to say, the rule in Mr. Darwin's speculations as to man's genealogy. He carries that genealogy back to some ancient form of animal life somewhat like an existing larval Ascidian ; and he does this on the strength of the observations of Kowalevsky and Kuppfer. He assumes at once that the similarities of structure which those observers detected are due to descent instead of to independent similarity of evolution, though the latter mode of origin is at least possible,* and can hardly be considered improbable when we reflect on the close similarity independently induced in the eyes of fishes and cephalopods.

---

* See Professor Rolleston's ' Address at the Liverpool Meeting of the British Association, 1870.'

Quite

Quite recently, however, observations have been published by Dr. Donitz,* which render it necessary, at the least, to pause and reconsider the question before admitting the Ascidian ancestry of the Vertebrate sub-kingdom.

We now come to the consideration of a subject of great importance—namely, that of man's mental powers. Are they, as Mr. Darwin again and again affirms that they are,† different only in degree and not in kind from the mental powers of brutes? As is so often the case in discussions, the error to be combated is an implied negation. Mr. Darwin implies and seems to assume that when two things have certain characters in common there can be no fundamental difference between them.

To avoid ambiguity and obscurity, it may be well here to state plainly certain very elementary matters. The ordinary antecedents and concomitants of distinctly felt sensations may exist, with all their physical consequences, in the total absence of intellectual cognizance, as is shown by the well-known fact, that when through fracture of the spine the lower limbs of a man are utterly deprived of the power of feeling, the foot may nevertheless withdraw itself from tickling just as if a sensation was consciously felt. Amongst lower animals, a decapitated frog will join its hind feet together to push away an irritating object just as an uninjured animal will do. Here we have coadjusted actions resulting from stimuli which normally produce sensation, but occurring under conditions in which cerebral action does not take place. Did it take place we should have sensations, but by no means necessarily intellectual action.

'Sensation' is not 'thought,' and no amount of the former would constitute the most rudimentary condition of the latter, though sensations supply the conditions for the existence of 'thought' or 'knowledge.'

Altogether, we may clearly distinguish at least six kinds of action to which the nervous system ministers :—

I. That in which impressions received result in appropriate movements without the intervention of sensation or thought, as in the cases of injury above given. (This is the reflex action of the nervous system.)

II. That in which stimuli from without result in sensations through the agency of which their due effects are wrought out. (Sensation.)

---

* See 'Journal für Anatomie und Physiologie,' edited by Reichert and Dubois. Berlin.

† 'There is no fundamental difference between man and the higher mammals in their mental faculties.'—*Descent of Man*, vol. i. p. 35.

III.

III. That in which impressions received result in sensations which give rise to the observation of sensible objects.—Sensible perception.

IV. That in which sensations and perceptions continue to coalesce, agglutinate, and combine in more or less complex aggregations, according to the laws of the association of sensible perceptions.—Association.

The above four groups contain only indeliberate operations, consisting, as they do at the best, but of mere *presentative* sensible ideas in no way implying any reflective or *representative* faculty. Such actions minister to and form *Instinct.* Besides these, we may distinguish two other kinds of mental action, namely :—

V. That in which sensations and sensible perceptions are reflected on by thought and recognised as our own and we ourselves recognised by ourselves as affected and perceiving.—Self-consciousness.

VI. That in which we reflect upon our sensations or perceptions, and ask what they are and why they are.—Reason.

These two latter kinds of action are deliberate operations, performed, as they are, by means of representative ideas implying the use of a *reflective representative* faculty. Such actions distinguish the *intellect* or rational faculty. Now, we assert that possession in perfection of all the first four (*presentative*) kinds of action by no means implies the possession of the last two (*representative*) kinds. All persons, we think, must admit the truth of the following proposition :—

Two faculties are distinct, not in degree but *in kind*, if we may possess the one in perfection without that fact implying that we possess the other also. Still more will this be the case if the two faculties tend to increase in an inverse ratio. Yet this is the distinction between the *instinctive* and the *intellectual* parts of man's nature.

As to animals, we fully admit that they may possess all the first four groups of actions—that they may have, so to speak, mental images of sensible objects combined in all degrees of complexity, as governed by the laws of association. We deny to them, on the other hand, the possession of the last two kinds of mental action. We deny them, that is, the power of reflecting on their own existence or of enquiring into the nature of objects and their causes. We deny that they know that they know or know themselves in knowing. In other words, we deny them *reason.* The possession of the presentative faculty, as above explained, in no way implies that of the reflective faculty; nor does any amount of direct operation imply the power of asking the reflective question before mentioned, as to 'what' and 'why.'

According

According to our definition, then, given above, the faculties of men and those of other animals differ in kind ; and brutes low in the scale supply us with a good example in support of this distinctness ; for it is in animals generally admitted to be wanting in reason—such as insects (*e. g.* the ant and the bee)—that we have the very summit and perfection of instinct made known to us.

We will shortly examine Mr. Darwin's arguments, and see if he can bring forward a single instance of brute action implying the existence in it of the representative reflective power. Before doing so, however, one or two points as to the conditions of the controversy must be noticed.

In the first place, the position which we maintain is the one in possession—that which is commended to us by our intuitions, by ethical considerations, and by religious teaching universally. The *onus probandi* should surely therefore rest with him who, attacking the accepted position, maintains the essential similarity and fundamental identity of powers the effects of which are so glaringly diverse. Yet Mr. Darwin quietly assumes the whole point in dispute, by asserting identity of *intuition* where there is identity of *sensation* (vol. i. p. 36), which, of course, implies that there is no mental power whatever except sensation. For if the existence of another faculty were allowed by him, it is plain that the action of that other faculty might modify the effects of mere sensation in any being possessed of such additional faculty.

Secondly, it must be remembered that it is a law in all reasoning that where known causes are sufficient to account for any phenomena we shall not gratuitously call in additional causes. If, as we believe to be the case, there is no need whatever to call in the *representative* faculty as an explanation of brute mental action ;—if the phenomena brutes exhibit can be accounted for by the *presentative* faculty—that is, by the presence of sensible perceptions and emotions together with the reflex and co-ordinating powers of the nervous system ;—then to ascribe to them the possession of reason is thoroughly gratuitous.

Thirdly, in addition to the argument that brutes have not intellect because their actions can be accounted for without the exercise of that faculty, we have other and positive arguments in opposition to Mr. Darwin's view of their mental powers. These arguments are based upon the absence in brutes of articulate and rational speech, of true concerted action and of educability, in the human sense of the word. We have besides, what may be called an experimental proof in the same direction. For if the germs of a rational nature existed in brutes, such germs would certainly ere this have so developed as to
have

have produced unmistakeably rational phenomena, considering the prodigious lapse of time passed since the entombment of the earliest known fossils. To this question we will return later.

We shall perhaps be met by the assertion that many men may also be taken to be irrational animals, so little do the phenomena they exhibit exceed in dignity and importance the phenomena presented by certain brutes. But, in reply, it is to be remarked that we can only consider men who are truly men—not idiots, and that all *men*, however degraded their social condition, have self-consciousness properly so called, possess the gift of articulate and rational speech, are capable of true concerted action, and have a perception of the existence of right and wrong. On the other hand, no brute has the faculty of articulate, rational speech: most persons will also admit that brutes are not capable of truly concerted action, and we contend most confidently that they have no self-consciousness, properly so called, and no perception of the difference between truth and falsehood and right and wrong.

Let us now consider Mr. Darwin's facts in favour of an opposite conclusion.

1st. His testimony drawn from his own experience and information regarding the lowest races of men.

2nd. The anecdotes he narrates in favour of the intelligence of brutes.

In the first place, we have to thank our author for very distinct and unqualified statements as to the substantial unity of men's mental powers. Thus he tells us:—

'The Fuegians rank amongst the lowest barbarians; but I was continually struck with surprise how closely the three natives on board H. M. S. "Beagle," who had lived some years in England, and could talk a little English, resembled us in disposition, and in most of our mental qualities.'—vol. i. p. 34.

Again he adds:—

'The American aborigines, Negroes and Europeans differ as much from each other in mind as any three races that can be named; yet I was incessantly struck, whilst living with the Fuegians on board the "Beagle," with the many little traits of character, showing how similar their minds were to ours; and so it was with a full-blooded negro with whom I happened once to be intimate.'—vol. i. p. 232.

Again:—'Differences of this kind (mental) between the highest men of the highest races and the lowest savages, are connected by the finest gradations' (vol. i. p. 35).

Mr. Darwin, then, plainly tells us that all the essential mental characters of civilised man are found in the very lowest races
of

of men, though in a less completely developed state ; while, in comparing their mental powers with those of brutes, he says 'No doubt the difference in this respect is enormous' (vol. i. p. 34). As if, however, to diminish the force of this admission, he remarks, what no one would dream of disputing, that there are psychical phenomena common to men and to other animals. He says of man that

'He uses in common with the lower animals inarticulate cries to express his meaning, aided by gestures and the movements of the muscles of the face. This especially holds good with the more simple and vivid feelings, which are *but little connected with the higher intelligence.* Our cries of pain, fear, surprise, anger, together with their appropriate actions, and the murmur of a mother to her beloved child, are more expressive than any words.'—vol. i. p. 54.

But, inasmuch as it is admitted on all hands that man *is* an animal, and therefore has all the four lower faculties enumerated in our list, as well as the two higher ones, the fact that he makes use of common instinctive actions in no way diminishes the force of the distinction between him and brutes as regards the representative, reflective faculties. It rather follows as a matter of course from his animality that he should manifest phenomena common to him and to brutes. That man has a common nature with them is perfectly compatible with his having, besides, a superior nature and faculties of which no brute has any rudiment or vestige. Indeed, all the arguments and objections in Mr. Darwin's second chapter may be met by the fact that man being an animal, has corresponding faculties, whence arises a certain external conformity with other animals as to the modes of expressing some mental modifications. In the overlooking of this possibility of coexistence of two natures lies that error of negation to which we before alluded. Here, as in other parts of the book, we may say there are two quantities a and $a + x$, and Mr. Darwin, seeing the two as but neglecting the x, represents the quantities as equal.

We will now notice the anecdotes narrated by Mr. Darwin in support of the rationality of brutes. Before doing so, however, we must remark that our author's statements, given on the authority (sometimes second-hand authority) of others, afford little evidence of careful criticism. This is the more noteworthy when we consider the conscientious care and pains which he bestows on all the phenomena which he examines himself.

Thus, for example, we are told on the authority of Brehm that—

'An eagle seized a young cercopithecus, which, by clinging to a branch, was not at once carried off; it cried loudly for assistance, upon

upon which other members of the troop, with much uproar rushed to the rescue, surrounded the eagle, and pulled out so many feathers that he no longer thought of his prey, but only how to escape.'—vol. i. p. 76.

We confess we wish that Mr. Darwin had himself witnessed this episode. Perhaps, however, he has seen other facts sufficiently similar to render this one credible. In the absence of really trustworthy evidence we should, however, be inclined to doubt the fact of a young cercopithecus, unexpectedly seized, being able, by clinging, to resist the action of an eagle's wings.

We are surprised that Mr. Darwin should have accepted the following tale without suspicion :—

'One female baboon had so capacious a heart that she not only adopted young monkeys of other species, but stole young dogs and cats which she continually carried about. Her kindness, however, did not go so far as to share her food with her adopted offspring, at which Brehm was surprised, as his monkeys always divided everything quite fairly with their own young ones. An adopted kitten scratched the above-mentioned affectionate baboon, *who certainly had a fine intellect,* for she was much astonished at being scratched, and immediately examined the kitten's feet, and without more ado bit off the claws.' (!!) —vol. i. p. 41.

Has Mr. Darwin ever tested this alleged fact? Would it be possible for a baboon to bite off the claws of a kitten without keeping the feet perfectly straight?

Again we have an anecdote on only second-hand authority (namely a quotation by Brehm of Schimper) to the following effect :—

'In Abyssinia, when the baboons belonging to one species (*C. gelada*) descend in troops from the mountains to plunder the fields, they sometimes encounter troops of another species (*C. hamadryas*), and then a fight ensues. The Geladas roll down great stones, which the Hamadryas try to avoid, and then both species, making a great uproar, rush furiously against each other. Brehm, when accompanying the Duke of Coburg-Gotha, aided in an attack with fire-arms on a troop of baboons in the pass of Mensa in Abyssinia. The baboons in return rolled so many stones down the mountain, some as large as a man's head, that the attackers had to beat a hasty retreat; and the pass was actually for a time closed against the caravan. It deserves notice that these baboons thus acted in concert.'—vol. i. p. 51.

Now, if every statement of fact here given be absolutely correct, it in no way even tends to invalidate the distinction we have drawn between 'instinct' and 'reason'; but the positive assertion that the brutes 'acted in concert,' when the evidence proves nothing more than that their actions were simultaneous, shows

shows a strong bias on the part of the narrator. A flock of sheep will simultaneously turn round and stare and stamp at an intruder; but this is not 'concerted action,' which means that actions are not only simultaneous, but are so in consequence of a reciprocal understanding and convention between the various agents. It may be added that if any brutes were capable of such really *concerted* action, the effects would soon make themselves known to us so forcibly as to prevent the possibility of mistake.

We come now to Mr. Darwin's instances of brute rationality. In the first place he tells us :—

'I had a dog who was savage and averse to all strangers, and I purposely tried his memory after an absence of five years and two days. I went near the stable where he lived, and shouted to him in my old manner; he showed no joy, but instantly followed me out walking and obeyed me, exactly as if I had parted with him only half an hour before. A train of old associations, dormant during five years, had thus been instantaneously awakened in his mind.'—vol. i. p. 45.

No doubt! but this is not 'reason.' Indeed, we could hardly have a better instance of the mere action of associated sensible impressions. What is there here which implies more than memory, impressions of sensible objects and their association? Had there been reason there would have been signs of joy and wonder, though such signs would not alone prove reason to exist. It is evident that Mr. Darwin's own mode of explanation is the sufficient one—namely, by a train of associated sensible impressions. Mr. Darwin surely cannot think that there is in this case any evidence of the dog's having put to himself those questions which, under the circumstances, a rational being would put. Mr. Darwin also tells us how a monkey-trainer gave up in despair the education of monkeys, of which the attention was easily distracted from his teaching, while 'a monkey which carefully attended to him could always be trained.' But 'attention' does not imply 'reason.' The anecdote only shows that some monkeys are more easily impressed and more retentive of impressions than others.

Again, we are told, as an instance of *reason*, that 'Rengger sometimes put a live wasp in paper so that the monkeys in hastily unfolding it got stung; after this had once happened, they always first held the packet to their ears to detect any movement within.' But here again we have no need to call in the aid of 'reason.' The monkeys had had the group of sensations 'folded paper' associated with the other groups—'noise and movement' and 'stung fingers.' The second time they experience

ence the group of sensations 'folded paper' the succeeding sensations (in this instance only too keenly associated) are forcibly recalled, and with the recollection of the sensation of hearing, the hand goes to the ear. Yet Mr. Darwin considers this unimportant instance of such significance that he goes on to say :—

'Any one who is not convinced by such facts as these, and by what he may observe with his own dogs, that animals can reason, would not be convinced by anything I could add. Nevertheless, I will give one case with respect to dogs, as it rests on two distinct observers, and can *hardly depend on the modification of any instinct.* Mr. Colquhoun winged two wild ducks, which fell on the opposite side of a stream ; his retriever tried to bring over both at once, but could not succeed ; she then, though never before known to ruffle a feather, deliberately killed one, brought over the other, and returned for the dead bird. Colonel Hutchinson relates that two partridges were shot at once, one being killed and the other wounded ; the latter ran away, and was caught by the retriever, who on her return came across the dead bird ; she stopped, evidently greatly puzzled, and after one or two trials, finding she could not take it up without permitting the escape of the winged bird, she considered a moment, then deliberately murdered it by giving it a severe crunch, and afterwards brought away both together. This was the only known instance of her having wilfully injured any game.'

Mr. Darwin adds :

'Here we have reason, though not quite perfect, for the retriever might have brought the wounded bird first and then returned for the dead one, as in the case of the two wild ducks.'—vol. i. pp. 47, 48.

Here we reply we have nothing of the kind, and to bring 'reason' into play is gratuitous. The circumstances can be perfectly explained (and on Mr. Darwin's own principles) as evidences of the revival of an old instinct. The ancestors of sporting dogs of course killed their prey, and that trained dogs do not kill it is simply due to man's action, which has suppressed the instinct by education, and which continually thus keeps it under control. It is indubitable that the old tendency *must* be latent, and that a small interruption in the normal retrieving process, such as occurred in the cases cited, would probably be sufficient to revive that old tendency and call the obsolete habit into exercise.

But perhaps the most surprising instance of groundless inference is presented in the following passage :—

'My dog, a full grown and very sensible animal, was lying on the lawn during a hot and still day ; but at a little distance a slight breeze occasionally moved an open parasol, which would have been wholly disregarded by the dog, had any one stood near it. As it was,
every

every time that the parasol slightly moved, the dog growled fiercely and barked. He must, I think, have reasoned to himself in a rapid and unconscious manner, that movement without any apparent cause indicated the presence of some strange living agent, and no stranger had a right to be on his territory.'—vol. i. p. 67.

The consequences deduced from this trivial incident are amazing. Probably, however, Mr. Darwin does not mean what he says; but, on the face of it, we have a brute credited with the abstract ideas 'movement,' 'causation,' and the notions logically arranged and classified in subordinate genera—'agent,' 'living agent,' 'strange living agent.' He also attributes to it the notion of 'a right' of 'territorial limitation,' and the relation of such 'limited territory' and 'personal ownership.' It may safely be affirmed that if a dog could so reason in one instance he would in others, and would give much more unequivocal proofs for Mr. Darwin to bring forward.

Mr. Darwin, however, speaks of reasoning in an 'unconscious manner,' so that he cannot really mean any process of reasoning at all; but, if so, his case is in no way apposite. Even an insect can be startled, and will exhibit as much evidence of rationality as is afforded by the growl of a dog; and all that is really necessary to explain such a phenomenon exists in an oyster, or even in the much talked-of Ascidian.

Thus, then, it appears that, even in Mr. Darwin's specially-selected instances, there is not a tittle of evidence tending, however slightly, to show that any brute possesses the representative reflective faculties. But if, as we assert, brute animals are destitute of such higher faculties, it may well be that those lower faculties which they have (and which we more or less share with them) are highly developed, and their senses possess a degree of keenness and quickness inconceivable to us. Their minds*
being entirely occupied with such lower faculties, and having, so to speak, nothing else to occupy them, their sensible impressions may become interwoven and connected to a far greater extent than in us. Indeed, in the absence of free will, the laws of this association of ideas obtain supreme command over the minds of brutes: the brute being entirely immersed, as it were, in his presentative faculties.

There yet remain two matters for consideration, which tend to prove the fundamental difference which exists between the mental powers of man and brutes:—1. The mental equality

* The words 'mind,' 'mental,' 'intelligence,' &c., are here made use of in reference to the psychical faculties of brutes, in conformity to popular usage, and not as strictly appropriate.

between

between animals of very different grades of structure, and their non-progressiveness ; 2. The question of articulate speech.

Considering the vast antiquity of the great animal groups,[*] it is, indeed, remarkable how little advance in mental capacity has been achieved even by the highest brutes. This is made especially evident by Mr. Darwin's own assertions as to the capacities of lowly animals. Thus he tells us that—

' Mr. Gardner, whilst watching a shore-crab (Gelasimus) making its burrow, threw some shells towards the hole. One rolled in, and three other shells remained within a few inches of the mouth. In about five minutes the crab brought out the shell which had fallen in, and carried it away to the distance of a foot ; it then saw the three other shells lying near, and *evidently thinking* that they might likewise roll in, carried them to the spot where it had laid the first.'— vol. i. p. 334.

Mr. Darwin adds or quotes the astonishing remark, ' It would, I think, be difficult to distinguish this act from one performed by man by the aid of reason.' Again, he tells us :—

' Mr. Lonsdale informs me that he placed a pair of land-shells (*Helix pomatia*), one of which was weakly, into a small and ill-provided garden. After a short time the strong and healthy individual disappeared, and was traced by its track of slime over a wall into an adjoining well-stocked garden. Mr. Lonsdale concluded that it had deserted its sickly mate ; but after an absence of twenty-four hours it returned, and apparently communicated the result of its successful exploration, for both then started along the same track and disappeared over the wall.'—vol. i. p. 325.

Whatever may be the real value of the statements quoted, they harmonize with a matter which is incontestable. We refer to the fact that the intelligence of brutes, be they high or be they low, is essentially one in kind, there being a singular parity between animals belonging to groups widely different in type of structure and in degree of development.

Apart from the small modifications which experience occasionally introduces into the habits of animals—as sometimes occurs after man has begun to frequent a newly-discovered island —it cannot be denied that, looking broadly over the whole animal kingdom, there is no evidence of advance in mental power on the part of brutes. This absence of progression in animal intelligence is a very important consideration, and it is one which does not seem to be adverted to by Mr. Darwin,

---

[*] Mr. Darwin (vol. i. p. 360) refers to Dr. Scudder's discovery of ' a fossil insect in the Devonian formation of New Brunswick, furnished with the well-known tympanum or stridulating apparatus of the male Locustidæ.'

though

though the facts detailed by him are exceedingly suggestive of it.

When we speak of this absence of progression we do not, of course, mean to deny that the dog is superior in mental activity to the fish, or the jackdaw to the toad. But we mean that, considering the vast period of time that must (on Mr. Darwin's theory) have elapsed for the evolution of an Orang from an Ascidian, and considering how beneficial increased intelligence must be to all in the struggle for life, it is inconceivable (on Mr. Darwin's principles only) that a mental advance should not have taken place greater in degree, more generally diffused, and more in proportion to the grade of the various animals than we find to be actually the case. For in what respect is the intelligence of the ape superior to that of the dog or of the elephant? It cannot be said that there is one point in which its psychical nature approximates to man more than that of those four-footed beasts. But, again, where is the great superiority of a dog or an ape over a bird? The falcon trained to hawking is at least as remarkable an instance of the power of education as the trained dog. The tricks which birds can be taught to perform are as complex and wonderful as those acted by the mammal. The phenomena of nidification, and some of those now brought forward by Mr. Darwin as to courtship, are fully comparable with analogous phenomena of quasi-intelligence in any beast.

This, however, is but a small part of the argument. For let us descend to the invertebrata, and what do we find?—a restriction of their quasi-mental faculties proportioned to their constantly inferior type of structure? By no means. We find, *e. g.*, in ants, phenomena which simulate those of an intelligence such as ours far more than do any phenomena exhibited by the highest beasts. Ants display a complete and complex political organization, classes of beings socially distinct, war resulting in the capture of slaves, and the appropriation and maintenance of domestic animals (*Aphides*) analogous to our milk-giving cattle.

Mr. Darwin truthfully remarks on the great difference in these respects between such creatures as ants and bees, and singularly inert members of the same class—such as the scale insect or coccus. But can it be pretended that the action of natural and sexual selection has alone produced these phenomena in certain insects, and failed to produce them in any other mere animals even of the very highest class? If these phenomena are due to a power and faculty similar in kind to human intelligence, and which power is latent and capable of evolution in all animals, then it is certain that this power must have been evolved in other instances also, and that we should see varying degrees of it in
many

many, and notably in the highest brutes as well as in man. If, on the other hand, the faculties of brutes are different in kind from human intelligence, there can be no reason whatever why animals most closely approaching man in physical structure should resemble him in psychical nature also.

This reflection leads us to the difference which exists between men and brutes as regards the faculty of articulate speech.   Mr. Darwin remarks that of the distinctively human characters this has 'justly been considered as one of the chief' (vol. i. p. 53). We cannot agree in this.   Some brutes can articulate, and it is quite conceivable that brutes might (though as a fact they do not) so associate certain sensations and gratifications with certain articulate sounds as, in a certain sense, to speak.   This, however, would in no way even tend to bridge over the gulf which exists between the representative reflective faculties and the merely presentative ones.   Articulate signs of sensible impressions would be fundamentally as distinct as mere gestures are from truly rational speech.

Mr. Darwin evades the question about language by in one place (vol. i. p. 54) attributing that faculty in man to his having acquired a higher intellectual nature ; and in another (vol. ii. p. 391), by ascribing his higher intellectual nature to his having acquired that faculty.

Our author's attempts to bridge over the chasm which separates instinctive cries from rational speech are remarkable examples of groundless speculation.   Thus he ventures to say—

'That primeval man, or rather some early progenitor of man, *probably* used his voice largely, as does one of the gibbon-apes at the present day, in producing true musical cadences, that is in singing ; we may conclude from a widely-spread analogy that this power would have been especially exerted during the courtship of the sexes, serving to express various emotions, as love, jealousy, triumph, and serving as a challenge to their rivals.   The imitation by articulate sounds of musical cries *might* have given rise to words expressive of various complex emotions.'

And again :

'It does not appear *altogether incredible*, that some unusually wise ape-like animal should have thought of imitating the growl of a beast of prey, so as to indicate to his fellow monkeys the nature of the expected danger.   And this would have been a first step in the formation of a language.'—vol. i. p. 56.

But the question, not whether it is incredible, but whether there are any data whatever to warrant such a supposition. Mr. Darwin brings forward none : we suspect none could be brought forward.

It

It is not, however, emotional expressions or manifestations of sensible impressions, in whatever way exhibited, which have to be accounted for, but the enunciation of distinct deliberate judgments as to 'the what,' 'the how,' and 'the why,' by definite articulate sounds; and for these Mr. Darwin not only does not account, but he does not adduce anything even tending to account for them. Altogether we may fairly conclude, from the complete failure of Mr. Darwin to establish identity of kind between the mental faculties of man and of brutes, that identity cannot be established; as we are not likely for many years to meet with a naturalist so competent to collect and marshal facts in support of such identity, if any such facts there are. The old barrier, then, between 'presentative instinct' and 'representative reason' remains still unimpaired, and, as we believe, insurmountable.

We now pass to another question, which is of even greater consequence than that of man's intellectual powers. Mr. Darwin does not hesitate to declare that even the 'moral sense' is a mere result of the development of brutal instincts. He maintains, 'the first foundation or origin of the moral sense lies in the social instincts, including sympathy; and these instincts no doubt were primarily gained, as in the case of the lower animals, through natural selection' (vol. ii. p. 394).

Everything, however, depends upon what we mean by the 'moral sense.' It is a patent fact that there does exist a perception of the qualities 'right' and 'wrong' attaching to certain actions. However arising, men have a consciousness of an absolute and immutable rule *legitimately* claiming obedience with an authority necessarily supreme and absolute—in other words, intellectual judgments are formed which imply the existence of an ethical ideal in the judging mind.

It is the existence of this power which has to be accounted for; neither its application nor even its validity have to be considered. Yet instances of difference of opinion respecting the moral value of particular concrete actions are often brought forward as if they could disprove the *existence* of moral intuition. Such instances are utterly beside the question. It is amply sufficient for our purpose if it be conceded that developed reason dictates to us that certain modes of action, abstractedly considered, are intrinsically wrong; and this we believe to be indisputable.

It is equally beside the question to show that the existence of mutually beneficial acts and of altruistic habits can be explained by 'natural selection.' No amount of benevolent habits tend even in the remotest degree to account for the intellectual perception

ception

ception of 'right' and 'duty.' Such habits may make the doing of beneficial acts pleasant, and their omission painful; but such feelings have essentially nothing whatever to do with the perception of 'right' and 'wrong,' nor will the faintest incipient stage of the perception be accounted for by the strongest development of such sympathetic feelings. Liking to do acts which happen to be good, is one thing; seeing that actions are good, whether we or others like them or not, is quite another.

Mr. Darwin's account of the moral sense is very different from the above. It may be expressed most briefly by saying that it is the prevalence of more enduring instincts over less persistent ones —the former being social instincts, the latter personal ones. He tells us:—

'As man cannot prevent old impressions continually repassing through his mind, he will be compelled to compare the weaker impressions of, for instance, past hunger, or of vengeance satisfied or danger avoided at the cost of other men, with the instinct of sympathy and goodwill to his fellows, which is still present and ever in some degree active in his mind. He will then feel in his imagination that a stronger instinct has yielded to one which now seems comparatively weak; and then that sense of dissatisfaction will inevitably be felt with which man is endowed, like every other animal, in order that his instincts may be obeyed.'—vol. i. p. 90.

Mr. Darwin means by 'the moral sense' an instinct, and adds, truly enough, that 'the very essence of an instinct is, that it is followed independently of reason' (vol. i. p. 100). But the very essence of moral action is that it is *not* followed independently of reason.

Having stated our wide divergence from Mr. Darwin with respect to what the term 'moral sense' denotes, we might be dispensed from criticising instances which must from our point of view be irrelevant, as Mr. Darwin would probably admit. Nevertheless, let us examine a few of these instances, and see if we can discover in them any justification of the views he propounds.

As illustrations of the development of self-reproach for the neglect of some good action, he observes:—

'A young pointer, when it first scents game, apparently cannot help pointing. A squirrel in a cage who pats the nuts which it cannot eat, as if to bury them in the ground, can hardly be thought to act thus either from pleasure or pain. Hence the common assumption that men must be impelled to every action by experiencing some pleasure or pain may be erroneous. Although a habit may be blindly and implicitly followed, independently of any pleasure or pain felt at the moment, yet if it be forcibly and abruptly checked, a vague sense of of dissatisfaction is generally experienced; and this is especially true in regard to persons of feeble intellect.'—vol. i. p. 80.

Now, passing over the question whether in the ' pointing' and ' patting' referred to there may not be some agreeable sensations, we contend that such instincts have nothing to do with 'morality,' from their blind nature, such blindness simply *ipso facto* eliminating every vestige of morality from an action.

Mr. Darwin certainly exaggerates the force and extent of social sympathetic feelings. Mr. Mill admits that they are ' often wanting;' but Mr. Darwin claims the conscious possession of such feelings for all, and quotes Hume as saying that the view of the happiness of others ' communicates a secret joy,' while the appearance of their misery ' throws a melancholy damp over the imagination.'* One might wish that this remark were universally true, but unfortunately some men take pleasure in the pain of others; and Larochefoucauld even ventured on the now well-known saying, ' that there is something in the misfortunes of our best friends not unpleasant to us.' But our feeling that the sufferings of others are pleasant or unpleasant has nothing to do with the question, which refers to the *judgment* whether the indulging of such feelings is ' right' or ' wrong.'

If the ' social instinct' were the real basis of the moral sense, the fact that society approved of anything would be recognised as the supreme sanction of it. Not only, however, is this not so, not only do we judge as to whether society in certain cases is right or wrong, but we demand a reason why we should obey society at all; we demand a rational basis and justification for social claims, if we happen to have a somewhat inquiring turn of mind. We shall be sure avowedly or secretly to despise and neglect the performance of acts which we do not happen to desire, and which have not an intellectual sanction.

The only passage in which our author seems as if about to meet the real question at issue is very disappointing, as the difficulty is merely evaded. He remarks, ' I am aware that some persons maintain that actions performed impulsively do not come under the dominion of the moral sense, and cannot be called moral' (vol. i. p. 87). This is not a correct statement of the intuitive view, and the difficulty is evaded thus: ' But it appears scarcely possible to draw any clear line of distinction of this kind, though the distinction may be real!' It seems to us, however, that there is no difficulty at all in drawing a line

* ' Enquiry concerning the Principles of Morals,' Edit. 1751, p. 132.

between

between a judgment as to an action being right or wrong and every other kind of mental act. Mr. Darwin goes on to say :—

'Moreover, an action repeatedly performed by us, will at last be done without deliberation or hesitation, and can then hardly be distinguished from an instinct; yet surely no one will pretend that an action thus done ceases to be moral. On the contrary, we all feel that an act cannot be considered as perfect, or as performed in the most noble manner, unless it is done impulsively, without deliberation or effort, in the same manner as by a man in whom the requisite qualities are innate.'—vol. i. p. 88.

To this must be replied, in one sense, 'Yes;' in another, 'No.' An action which has ceased to be directly or indirectly deliberate has ceased to be moral as a distinct act, but it is moral as the continuation of those preceding deliberate acts through which the good habit was originally formed, and the rapidity with which the will is directed in the case supposed may indicate the number and constancy of antecedent meritorious volitions. Mr. Darwin seems to see this more or less, as he adds: 'He who is forced to overcome his fear or want of sympathy before he acts, deserves, however, in one way higher credit than the man whose innate disposition leads him to a good act without effort.'

As an illustration of the genesis of remorse, we have the case

'of a temporary though for the time strongly persistent instinct conquering another instinct which is usually dominant over all others.' Swallows 'at the proper season seem all day long to be impressed with the desire to migrate; their habits change; they become restless, are noisy, and congregate in flocks. Whilst the mother-bird is feeding or brooding over her nestlings, the maternal instinct is probably stronger than the migratory; but the instinct which is more persistent gains the victory, and at last, at a moment when her young ones are not in sight, she takes flight and deserts them. When arrived at the end of her long journey, and the migratory instinct ceases to act, what an agony of remorse each bird would feel, if, from being endowed with great mental activity, she could not prevent the image continually passing before her mind of her young ones perishing in the bleak north from cold and hunger.'—vol. i. p. 90.

Let us suppose she does suffer 'agony,' that feeling would be nothing to the purpose. What is requisite is that she shall judge that she *ought not* to have left them. To make clear our point, let us imagine a man formerly entangled in ties of affection which in justice to another his conscience has induced him to sever. The image of the distress his act of severance has caused may occasion him keen emotional suffering for years, accompanied by a clear perception that his act has been right. Again, let us
suppose

suppose another case : The struggling father of a family becomes aware that the property on which he lives really belongs to another, and he relinquishes it. He may continue to judge that he has done a proper action, whilst tortured by the trials in which his act of justice has involved him. To assert that these acts are merely instinctive would be absurdly false. In the cases supposed, obedience is paid to a clear intellectual perception and against the very strongest instincts.

That we have not misrepresented Mr. Darwin's exposition of ' conscience' is manifest. He says that if a man has gratified a passing instinct, to the neglect of an enduring instinct, he ' will then feel dissatisfied with himself, and will resolve with more or less force to act differently for the future. This is conscience ; for conscience looks backwards and judges past actions, inducing that kind of dissatisfaction, which if weak we call regret, and if severe remorse' (vol. i. p. 91.) 'Conscience' certainly ' looks back and judges,' but not all that ' looks back and judges' is ' conscience.' A judgment of conscience is one of a particular kind, namely a judgment according to the standard of moral worth. But for this, a *gourmand*, looking back and judging that a particular sauce had occasioned him dyspepsia, would, in the dissatisfaction arising from his having eaten the wrong dish at dinner, exercise his conscience !

Indeed, elsewhere (vol. i. p. 103) Mr. Darwin speaks of ' the standard of morality rising higher and higher,' though he nowhere explains what he means either by the ' standard' or by the ' higher ;' and, indeed, it is very difficult to understand what can possibly be meant by this ' rising of the standard,' if the ' standard' is from first to last pleasure and profit.

We find, again, the singular remark :—' If any desire or instinct leading to an action opposed to the good of others, still appears to a man, when recalled to mind, as strong as or stronger than his social instinct, he will feel no keen regret at having followed it' (vol. i. p. 92).

Mr. Darwin is continually mistaking a merely beneficial action for a moral one ; but, as before said, it is one thing to *act well* and quite another to be a moral agent. A dog or even a fruit-tree may act well, but neither is a moral agent. Of course, all the instances he brings forward with regard to animals are not in point, on account of this misconception of the problem to be solved. He gives, however, some examples which tell strongly against his own view. Thus, he remarks of the *Law of Honour* —' The breach of this law, even when the breach is known to be strictly accordant with true morality, has caused many a man more agony than a real crime. We recognise the same influence in in the sense of burning shame which most of us have felt, even after the interval of years, when calling to mind some accidental breach of a trifling, though fixed, rule of etiquette' (vol. i. p. 92). This is most true; some trifling breach of good manners may indeed occasion us pain; but this may be unaccompanied by a judgment that we are morally blameworthy. It is judgment, and not feeling, which has to do with right and wrong. But a yet better example might be given. What quality can have been more universally useful to social communities than courage? It has always been, and is still, greatly admired and highly appreciated, and is especially adapted, both directly and indirectly, to enable its possessors to become the fathers of succeeding generations. If the social instinct were the basis of the moral sense, it is infallibly certain that courage must have come to be regarded as supremely 'good,' and cowardice to be deserving of the deepest moral condemnation. And yet what is the fact? A coward feels probably self-contempt and that he has incurred the contempt of his associates, but he does not feel 'wicked.' He is painfully conscious of his defective organization, but he knows that an organization, however defective, cannot, in itself, constitute moral demerit. Similarly, we, the observers, despise, avoid, or hate a coward; but we can clearly understand that a coward may be a more virtuous man than another who abounds in animal courage.

The better still to show how completely distinct are the conceptions 'enduring or strong instincts' and 'virtuous desires' on the one hand, and 'transient or weak impulses' and 'vicious inclinations' on the other, let us substitute in the following passage for the words which Mr. Darwin, on his own principles, illegitimately introduces, others which accord with those principles, and we shall see how such substitution eliminates every element of morality from the passage:—

'Looking to future generations, there is no cause to fear that the social instincts will grow weaker, and we may expect that enduring [virtuous] habits will grow stronger, becoming perhaps fixed by inheritance. In this case the struggle between our stronger [higher] and weaker [lower] impulses will be less severe, and the strong [virtue] will be triumphant' (vol. i. p. 104).

As to past generations, Mr. Darwin tells us (vol. i. p. 166) that at all times throughout the world tribes have supplanted other tribes; and as social acts are an element in their success, sociality must have been intensified, and this because 'an increase in the number of well-endowed men will certainly give an immense advantage to one tribe over another.' No doubt! but this only

only explains an augmentation of mutually beneficial actions. It does not in the least even tend to explain how the moral judgment was first formed.

Having thus examined Mr. Darwin's theory of Sexual Selection, and his comparison of the mental powers of man (including their moral application) with those of the lower animals, we have a few remarks to make upon his mode of conducting his argument.

In the first place we must repeat what we have already said as to his singular dogmatism, and in the second place we must complain of the way in which he positively affirms again and again the existence of the very things which have to be proved. Thus, to take for instance the theory of the descent of man from some inferior form, he says :—' the grounds upon which this conclusion rests *will never be shaken*' (vol. ii. p. 385), and ' the possession of exalted mental powers is *no* insuperable objection to this conclusion' (vol. i. p. 107). Speaking of sympathy, he boldly remarks, — ' this instinct *no doubt* was originally acquired like all the other social instincts through natural selection' (vol. i. p. 164); and ' the fundamental social instincts *were* originally thus gained' (vol. i. p. 173).

Again, as to the stridulating organs of insects, he says :—' No one who admits the agency of natural selection, will dispute that these musical instruments have been acquired through sexual selection.' Speaking of the peculiarities of humming-birds and pigeons, Mr. Darwin observes, ' the *sole* difference between these cases is, that in one the result is due to man's selection, whilst in the other, as with humming-birds, birds of paradise, &c., it is due to sexual selection,—that is, to the selection by the females of the more beautiful males' (vol. ii. p. 78.) Of birds, the males of which are brilliant, but the hens are only slightly so, he remarks : ' these cases *are almost certainly* due to characters primarily acquired by the male, having been transferred, in a greater or less degree, to the female' (vol. ii. p. 128). ' The colours of the males may *safely* be attributed to sexual selection' (vol. ii. p. 194). As to certain species of birds in which the males alone are black, we are told, there can *hardly be a doubt*, that blackness in these cases has been a sexually selected character' (vol. ii. p. 226). The following, again, is far too positive a statement:—' Other characters proper to the males of the lower animals, such as bright colours, and various ornaments *have been* acquired by the more attractive males having been preferred by the females. There are, however, exceptional cases, in which the males, instead of having been selected, *have been* the selectors' (vol. ii. p. 371).

It

It is very rarely that Mr. Darwin fails in courtesy to his opponents ; and we were therefore surprised at the tone of the following passage (vol. ii. p. 386) :—' He who is not content to look, *like a savage*, at the phenomena of nature as disconnected, *cannot* any longer believe that man is the work of a separate act of creation. He will be *forced* to admit' the contrary. What justifies Mr. Darwin in his assumption that to suppose the soul of man to have been specially created, is to regard the phenomena of nature as disconnected ?

In connexion with this assumption of superiority on Mr. Darwin's part, we may notice another matter of less importance, but which tends to produce the same effect on the minds of his readers. We allude to the terms of panegyric with which he introduces the names or opinions of every disciple of evolutionism, while writers of equal eminence, who have not adopted Mr. Darwin's views, are quoted, for the most part, without any commendation. Thus we read of our ' great anatomist and philosopher, Prof. Huxley,'—of ' our great philosopher, Herbert Spencer,'—of 'the remarkable work of Mr. Galton,'—of 'the admirable treatises of Sir Charles Lyell and Sir John Lubbock,' —and so on. We do not grudge these gentlemen such honorific mention, which some of them well deserve, but the repetition produces an unpleasant effect; and we venture to question the good taste on Mr. Darwin's part, in thus speaking of the adherents to his own views, when we do not remember, for example, a word of praise bestowed upon Prof. Owen in the numerous quotations which our author has made from his works.

Secondly, as an instance of Mr. Darwin's practice of begging the question at issue, we may quote the following assertion :— ' Any animal whatever, endowed with well-marked social *instincts*, would inevitably acquire a moral sense or *conscience*, as soon as its intellectual powers had become as well developed, or nearly as well developed, as in man' (vol. i. p. 71). This is either a monstrous assumption or a mere truism ; it is a truism, for of course, any creature with the intellect of a man would perceive the qualities men's intellect is capable of perceiving, and, amongst them—moral worth.

Mr. Darwin, in a passage before quoted (vol. i. p. 86) slips in the whole of absolute morality, by employing the phrase 'appreciation of justice.' Again (vol. i. p. 168), when he speaks of aiding the needy, he remarks :—' Nor could we check our sympathy, if so urged by hard reason, without deterioration in the *noblest* part of our nature.' How noblest ? According to Mr. Darwin, a virtuous instinct is a strong and permanent one. There can be, according to his views, no other elements of quality than

than intensity and duration. Mr. Darwin, in fact, thus silently and unconsciously introduces the moral element into his 'social instinct,' and then, of course, has no difficulty in finding in the latter what he had previously put there. This, however, is quite illegitimate, as he makes the social instinct synonymous with the gregariousness of brutes. In such gregariousness, however, there is no moral element, because the mental powers of brutes are not equal to forming reflective, deliberate, representative judgments.

The word 'social' is ambiguous, as gregarious animals may metaphorically be called social, and man's social relations may be regarded both beneficentially and morally. Having first used 'social' in the former sense, it is subsequently applied in the latter; and it is thus that the really moral conception is silently and illegitimately introduced.

We may now sum up our judgment of Mr. Darwin's work on the 'Descent of Man'—of its execution and tendency, of what it fails to accomplish and of what it has successfully attained.

Although the style of the work is, as we have said, fascinating, nevertheless we think that the author is somewhat encumbered with the multitude of his facts, which at times he seems hardly able to group and handle so effectively as might be expected from his special talent. Nor does he appear to have maturely reflected over the data he has so industriously collected. Moreover, we are surprised to find so accurate an observer receiving as facts many statements of a very questionable nature, as we have already pointed out, and frequently on second-hand authority. The reasoning also is inconclusive, the author having allowed himself constantly to be carried away by the warmth and fertility of his imagination. In fact, Mr. Darwin's power of reasoning seems to be in an inverse ratio to his power of observation. He now strangely exaggerates the action of 'sexual selection,' as previously he exaggerated the effects of the 'survival of the fittest.' On the whole, we are convinced that by the present work the cause of 'natural selection' has been rather injured than promoted; and we confess to a feeling of surprise that the case put before us is not stronger, since we had anticipated the production of far more telling and significant details from Mr. Darwin's biological treasure-house.

A great part of the work may be dismissed as beside the point—as a mere elaborate and profuse statement of the obvious fact, which no one denies, that man is an animal, and has all the essential properties of a highly organised one. Along with this truth, however, we find the assumption
that

that he is *no more* than an animal—an assumption which is necessarily implied in Mr. Darwin's distinct assertion that there is no difference of *kind*, but merely one of *degree*, between man's mental faculties and those of brutes.

We have endeavoured to show that this is distinctly untrue. We maintain that while there is no need to abandon the received position that man is truly an animal, he is yet the only rational one known to us, and that his rationality constitutes a fundamental distinction—one of *kind* and not of *degree*. The estimate we have formed of man's position differs therefore most widely from that of Mr. Darwin.

Mr. Darwin's remarks, before referred to (*ante*, p. 77), concerning the difference between the instincts of the coccus (or scale insect) and those of the ant—and the bearing of that difference on their zoological position (as both are members of the class insecta) and on that of man—exhibit clearly his misapprehension as to the true significance of man's mental powers.

For in the first place zoological classification is morphological. That is to say it is a classification based upon form and structure —upon the number and shape of the several parts of animals, and not at all upon what those parts *do*, the consideration of which belongs to physiology. This being the case we not only may, but *should*, in the field of zoology, neglect all questions of diversities of instinct or mental power, equally with every other power, as is evidenced by the location of the bat and the porpoise in the same class, mammalia, and the parrot and the tortoise in the same larger group, Sauropsida.

Looking, therefore, at man with regard to his bodily structure, we not only may, but *should*, reckon him as a member of the class mammalia, and even (we believe) consider him as the representative of a mere family of the first order of that class. But all men are not zoologists; and even zoologists must, outside their science, consider man in his totality and not merely from the point of view of anatomy.

If then we are right in our confident assertion that man's mental faculties are different *in kind* from those of brutes, and if he is, as we maintain, the only rational animal; then is man, as a whole, to be spoken of by preference from the point of view of his animality, or from the point of view of his rationality? Surely from the latter, and, if so, we must consider not structure, but action.

Now Mr. Darwin seems to concede* that a difference in kind *would* justify the placing of man in a distinct kingdom, inasmuch

* 'Descent of Man,' vol. i. p. 186.

as

as he says a difference in degree does not so justify; and we have no hesitation in affirming (with Mr. Darwin) that between the instinctive powers of the coccus and the ant there *is* but a difference of degree, and that, therefore, they do belong to the same kingdom; but we contend it is quite otherwise with man. Mr. Darwin doubtless admits that all the wonderful actions of ants are mere modifications of instinct. But if it were not so —if the piercing of tunnels beneath rivers, &c., were evidence of their possession of reason, then, far from agreeing with Mr. Darwin, we should say that ants also are rational animals, and that, while considered from the anatomical stand-point they would be insects, from that of their rationality they would rank together with man in a kingdom apart of 'rational animals.' Really, however, there is no tittle of evidence that ants possess the reflective, self-conscious, deliberate faculty; while the perfection of their instincts is a most powerful argument against the need of attributing a rudiment of rationality to any brute whatever.

We seem then to have Mr. Darwin on our side when we affirm that animals possessed of mental faculties distinct in kind should be placed in a kingdom apart. And man possesses such a distinction.

Is this, however, all that can be said for the dignity of his position? Is he merely one division of the visible universe co-ordinate with the animal, vegetable, and mineral kingdoms?

It would be so if he were intelligent and no more. If he could observe the facts of his own existence, investigate the co-existences and successions of phenomena, but all the time remain like the other parts of the visible universe a mere floating unit in the stream of time, incapable of one act of free self-determination or one voluntary moral aspiration after an ideal of absolute goodness. This, however, is far from being the case. Man is not merely an intellectual animal, but he is also a free moral agent, and, as such—and with the infinite future such freedom opens out before him—differs from all the rest of the visible universe by a distinction so profound that none of those which separate other visible beings is comparable with it. The gulf which lies between his being as a whole, and that of the highest brute, marks off vastly more than a mere kingdom of material beings; and man, so considered, differs far more from an elephant or a gorilla than do these from the dust of the earth on which they tread.

Thus, then, in our judgment the author of the 'Descent of Man' has utterly failed in the only part of his work which is really important. Mr. Darwin's errors are mainly due to a
radically

radically false metaphysical system in which he seems (like so many other physicists) to have become entangled. Without a sound philosophical basis, however, no satisfactory scientific superstructure can ever be reared ; and if Mr. Darwin's failure should lead to an increase of philosophic culture on the part of physicists, we may therein find some consolation for the injurious effects which his work is likely to produce on too many of our half-educated classes. We sincerely trust Mr. Darwin may yet live to furnish us with another work, which, while enriching physical science, shall not, with needless opposition, set at naught the first principles of both philosophy and religion.

19

Reprinted from pages 384–385 of *The Naturalist in Nicaragua: A Narrative of a Residence at the Gold Mines of Chontales; Journeys in the Savannahs and Forests. With Observations on Animals and Plants in Reference to the Theory of Evolution of Living Forms,* J. Murray, London, 1874, 401p.

MIMETIC INSECTS

T. Belt

[*Editor's Note:* In the original, material precedes and follows this excerpt.]

Viewed in this light, the wonderful perfection of mimetic forms is a natural consequence of the selection of the individuals that, on the one side, were more and more mimetic, and on the other (that of their enemies) more and more able to penetrate through the assumed disguises. It doubtless happened in some cases that species, having many foes, have entirely thrown off some of them through the disguises they have been brought to assume, but others they still cannot elude.

Since Mr. Bates first brought forward the theory of mimetic resemblances its importance has been more and more demonstrated, as it has been found how very largely animal life has been influenced in form and colour by the natural selection of the varieties that were preserved from their enemies, or enabled to approach their prey, through the resemblance they bore to something else. So general are these deceptive resemblances throughout nature, that it is often difficult to determine whether sexual preferences or the preservation of mimetic forms has been most potent in moulding the form and coloration of species, and in some the two forces are seen to be opposed in their operation. Thus in some butterflies that mimic the Heliconidæ, the females only are mimetic, the males retaining the normal form and coloration of the group to which they belong. In such cases it appears as if the females have not been checked in gradually assuming the disguise they wear, and it is important that they should be protected, as they are more exposed to destruction while seeking for places to deposit their eggs;

but that both sexes should not have inherited the change in form and colour when it would have been beneficial to both can only be explained, I think, on the supposition that the females had a choice of mates and preferred those that retained the primordial appearance of the group. This view is supported by the fact that many of the males of the mimetic *Leptalides* have the upper half of the lower wing of a pure white, whilst all the rest of the wings is barred and spotted with black, red, and yellow, like the species they mimic. The females have not this white patch, and the males usually conceal it by covering it with the upper wing, so that I cannot imagine its being of any other use to them than as an attraction in court-ship, when they exhibit it to the females, and thus gratify their deep-seated preference for the normal colour of the order to which the Leptalides belong.

20

Reprinted from *Zool. Soc. London Proc.* **25**:367–369 (1882)

2. On the Modification of a Race of Syrian Street-Dogs by means of Sexual Selection. By Dr. Van Dyck. With a Preliminary Notice by Charles Darwin, F.R.S., F.Z.S.

[Received April 4, 1882.]

Most of the naturalists who admit that natural selection has been effective in the formation of species, likewise admit that the weapons of male animals are the result of sexual selection—that is, of the best-armed males obtaining most females and transmitting their masculine superiority to their male offspring. But many naturalists doubt, or deny, that female animals ever exert any choice, so as to select certain males in preference to others. It would, however, be more correct to speak of the females as being excited or attracted in an especial degree by the appearance, voice, &c. of certain males, rather than of deliberately selecting them. I may perhaps be here permitted to say that, after having carefully weighed to the best of my ability the various arguments which have been advanced against the principle of sexual selection, I remain firmly convinced of its truth. It is, however, probable that I may have extended it too far, as, for instance, in the case of the strangely formed horns and mandibles of male Lamellicorn beetles, which have recently been discussed with much knowledge by W. von Reichenau[2], and about which I have always felt some doubts. On the other hand, the explanation of the development of the horns offered by this entomologist does not seem to me at all satisfactory.

[1] Annales des Sciences Naturelles, Zoologie, 6me série, tome viii. p. 6 (1879).
[2] " Ueber den Ursprung der secundären männlichen Geschlechtscharakteren &c.," Kosmos, Jahrgang v. 1881, p. 172.

In order to ascertain whether female animals ever or often exhibit a decided preference for certain males, I formerly inquired from some of the greatest breeders in England, who had no thoretical views to support and who had ample experience; and I have given their answers, as well as some published statements, in my 'Descent of Man'[1]. The facts there given clearly show that with dogs and other animals the females sometimes prefer in the most decided manner particular males—but that it is very rare that a male will not accept any female, though such cases do occur. The following statement, taken from the 'Voyage of the Vega,'[2] indirectly supports in a striking manner the above conclusion. Nordenskiöld says:—"We had two Scotch collies with us on the 'Vega.' They at first frightened the natives very much with their bark. To the dogs of the Chukches they soon took the same superior standing as the European claims for himself in relation to the savage. The dog was distinctly preferred by the female Chukch canine population, and that too without the fights to which such favour on the part of the fair commonly gives rise. A numerous canine progeny of mixed Scotch-Chukch breed has arisen at Pitlekay. The young dogs had a complete resemblance to their father; and the natives were quite charmed with them."

What the attractions may be which give an advantage to certain males in wooing in the above several cases, whether general appearance, such as colour and form, or vigour and strength, or gestures, voice, or odour, can rarely be even conjectured; but whatever they may be, they would be preserved and augmented in the course of many generations, if the females of the same species or race, inhabiting the same district, retained during successive generations approximately the same general disposition and taste; and this does not seem improbable. Nor is it indispensable that all the females should have exactly the same tastes: one female might be more attracted by some one characteristic in the male, and another female by a different one; and both, if not incompatible, would be gradually acquired by the males. Little as we can judge what are the characteristics which attract the female, yet, in some of the cases recorded by me, it seemed clearly to be colour; in other cases previous familiarity with a particular male; in others exactly the reverse, or novelty. With respect to the first appearance of the peculiarities which are afterwards augmented through sexual selection, this of course depends on the strong tendency in all parts of all organisms to present slight individual differences, and in some organisms to vary in a plain manner. Evidence has also been given in my book on Variation under Domestication showing that male animals are more liable to vary than females; and this would be highly favourable to sexual selection. Manifestly every slight individual difference and each more conspicuous variation depends on definite though unknown

[1] The Descent of Man, second edit. (1874), part ii. Chap. xvii. pp. 522–525. See also Chap. xiv., on choice in pairing shown by female birds, and on their appreciation of beauty.
[2] 'The Voyage of the Vega,' Eng. translat. (1881), vol. ii. p. 97.

causes; and these modifications of structure &c. differ in different species under apparently the same conditions. Statements of this nature have sometimes been misinterpreted, as if it were supposed that variations were indefinite or fluctuating, and that the same variations occurred in all species.

In reference to sexual selection, I will here only add that the complete manner in which the introduced dogs and other domestic animals in South America and other countries have been mongrelized, so that all traces of their original race have been lost, often appeared to me a surprising fact. This holds good according to Rengger[1] with the dogs even in so isolated a country as Paraguay. I formerly attributed this mongrelization merely to the breeds not having been kept separate and to the greater vigour of cross-bred offspring; but if the females often prefer strangers to their old companions, as seems to be the case, according to Nordenskiöld, in Siberia, and in Syria as shown in the following essay, then we can readily understand how rapid and complete would be the progress of mongrelization. I will now give without further comment the essay which Dr. W. Van Dyck, Lecturer on Zoology to the Protestant College at Beyrout, who has had excellent opportunities for observations during a residence of twenty years, has been so kind as to send me.

---

[1] 'Naturgeschichte der Säugethiere von Paraguay,' 1830, p. 154.

Part V

CONTROVERSY OVER SEXUAL SELECTION THEORY AFTER CHARLES DARWIN'S DEATH, 1882–1900

Editor's Comments
on Papers 21, 22, and 23

In 1889, seven years after Charles Darwin's death, two American biologists, George Peckham and Elizabeth Peckham, published their observations on the mating behavior of spiders belonging to the family Attidae (Paper 21). They concluded that their observations supported Darwin's theory that many differences in color and in ornamentation between the males and females of lower animals are the evolutionary outcome of female choice. They also used their observations to test the male vigor theory proposed by Wallace (1878; see also Paper 22) and concluded that their observations contradicted Wallace's theory.

Alfred Russel Wallace published an updated version of his views on natural selection and sexual selection in 1889 (Paper 22). He argued that sexual dimorphism in such characteristics as color, odor, and sound can be brought about by natural selection because of their value as a means of sexual recognition between the two sexes of a species (pp. 273, 284–285 in Paper 22). Wallace contended that female choice was ineffective because the

> extremely rigid action of natural selection must render any attempt to select mere ornament utterly nugatory, unless the most ornamented always coincide with "the fittest" in every other respect; while, if they do so coincide, then any selection of ornament is altogether superfluous. (p. 295 in Paper 22)

Wallace went on to argue, "The term 'sexual selection' must, therefore, be restricted to the direct results of male struggle and combat. This is really a form of natural selection. . . ." (p. 296 in Paper 22).

Wallace also contended that more intense coloring develops in "the male, due probably to his greater vigour and excitability" while "natural selection has caused the female to retain the primitive and more sober colours of the group for purposes of protection" (p. 298 in Paper 22). Geddes and Thomson contrasted the theories of Darwin and Wallace on female choice in their book *The Evolution of Sex* (1889, p. 29): "According to Darwin, sexual selection, for love's sake, has accelerated the males into gay colouring; according to Wallace, natural selection, for safety's sake, has retarded the females (birds or butterflies) and kept them inconspicuously plain." They advocated the following position:

> The true view seems to be, that both sexes have differentiated towards their respective goals, but males faster, because so katabolic; the limits are constantly being fixed by natural selection in Wallace's cases, and as constantly increased by sexual selection in Darwin's. There is, in fact, no reason why both should not be admitted as minor factors; but the greater part of the explanation is to be found in the view above stated, viz., in the physiological constitution of males and females themselves . . . the present position . . . regards gay colouring as the expression of the predominately katabolic or male sex, and quiet plainness as equally natural to the predominately anabolic females. (Geddes and Thomson, 1889, p. 29)

Peckham and Peckham (1890) published additional observations on sexual selection in 1890 and they used the opportunity to attack the updated version of the theories that Wallace was using in *Darwinism* (Paper 22) to explain the evolution of sexual dimorphism. The Peckhams' published still more observations of female choice in spiders nineteen years later (Peckham and Peckham, 1909, pp. 355–363). The Peckhams' assertions concerning the importance of female choice were criticized for many reasons (see, for example, Montgomery, 1910) that are no longer considered valid. Even the two potentially most damaging criticisms—that the Peckhams' conclusions were invalid because they made many of their observations under laboratory conditions with spiders in abnormally high densities and because females tend to copulate only once and with the first conspecific male that courts them in nature—have been shown to be invalid by the recent laboratory experiments conducted by Jackson (1980, 1981).

The research and conclusions of the Peckhams' were used by Poulton (1890) and Romanes (1892) to defend Charles Darwin's theory that female choice was one of the forces by which sexual selection brought about the evolution of sexual dimorphism in coloration and ornamentation. Poulton, however, later wrote:

> Probably the majority of naturalists are convinced by Darwin's arguments and his great array of facts that the principle of sexual

selection is real, and accounts for certain relatively unimportant features in the higher animals, and they further accept Darwin's opinion that its action has always been entirely subordinate to natural selection (Poulton, 1896, p. 188).

Darwin's ideas concerning female choice were applied by Rhoads (1890) in an attempt to explain the causes of polygyny in birds. Nutting (1891, p. 107), who surveyed sexual dimorphism in walruses, seals, and sea-lions, reported evidence for male-male combat but was "unable to find any evidence that the female Pinnipedia exercise any choice in the matter of accepting or rejecting individual successful males."

The role that female choice plays in evolution was reviewed by the comparative psychologists Morgan (1890, 1896, 1900) and Groos (1898). Morgan concluded that sexual selection, operating via preferential mating, may be a cause of sex differences in behavior "if it can be shown conclusively that choice is exercised, and that some are thereby excluded from mating" (Morgan, 1896, p. 230). Groos (1898, p. 244) tried to transform female choice into nothing but a special case of natural selection by contending that "the excited condition necessary for pairing, and also a certain difficulty in its execution, are both useful for the preservation of the species." Groos went on to conclude that the problem thus is no longer that of a conscious or unconscious choice but that of the male overcoming the reluctance of the female being woed.

Karl Pearson, an English mathematician who developed statistical methods in an attempt to quantitatively measure natural selection, contended:

> Natural selection requires selective mating, sexual selection in its broadest sense, to produce that barrier to intercrossing on which the origin of species depends. Darwin used sexual selection in the sense of the total or partial rejection of one type of mate by one or other sex. This I should prefer to term preferential mating. (Pearson, 1900, p. 423)

Pearson developed a mathematical model of his theory that speciation could occur sympatrically (without geographical isolation) via sexual selection. After contending that "with some form of sexual selection differentiation of a local race is a possibility" (Pearson, 1900, p. 441), he drew attention to the fact that all individuals who mate are not equally fertile. He did not attribute differential fertility among those who mate to sexual selection but rather considered this to be due to "reproductive selection," a term he had coined previously (Pearson, 1896) and that he considered to be distinct from natural selection (Pearson's "selective mortality") and sexual selection (Pearson's "preferential mating").

Darwin's discussion of the role that sexual selection has and is playing in human evolution centered around the evolution and maintenance of differences among human races. In France, Darwin's contention that sexual selection has played an important role in the evolution of human races was critically reviewed by Broca (1872) and defended by Sicard (1892). In Germany, Haeckel, one of the first scientists to champion Darwin's theory of sexual selection (Haeckel, 1866, vol. 2, pp. 244-247; 1868, 1874), supported Darwin's contention:

> Among men sexual selection has given rise to a noble form of family life, which is the chief foundation on which civilization and social states have been built. The human race certainly owes its origin in great measure to the perfected Sexual Selection which our ancestors exercised in the choice of wives (Haeckel, 1904, vol. 1, p. 103).

The Scandinavian sociologist Edward Westermarck (1891) published an extensive discussion of sexual selection theory in his book *The History of Human Marriage*. Westermarck (1891, pp. 542, 543) concluded that "sexual selection of lower animals is entirely subordinate to the great law of survival of the fittest"; that "there is an ideal of beauty common to the whole human race"; that "different standards of beauty are due to racial differences" in contrast to Darwin who held that racial differences were due to different standards of beauty; and that "physical beauty is therefore in every respect the outward manifestation of physical perfection, and the development of the instinct which prefers beauty to ugliness, healthiness to disease, is evidently within the power of natural selection."

Francis Galton, Thomas Huxley, and many other scholars attempted to ascertain the direction and intensity of selection with respect to human behavior in contemporary as well as past European populations. Both Galton and Huxley concluded that the economically and theologically based political practices of modern European societies were generating selection against those human beings who were intellectually and ethically fit, and that appeals to conscience—voluntarism—as a political solution to the problem only intensifies the selection against the intellectually and ethically fit (Bajema, 1976). It is interesting to note that Galton and Huxley rarely used the term "sexual selection" when they discussed the selection being generated by intraspecific competition for resources and mates both within and between human populations.

While Wallace opposed Darwin's theory that sexual selection played a major role in the evolution of human races in 1871 (Paper 17), and considered female choice to be an impotent force in the evolution of animal species (pp. 294-296 in Paper 22), he later expressed his belief that female choice could be an important eugenic force in future human evolution. Wallace advocated the establishment of a socialist society adapted to the equal well-being of all in which selection producing improvement in the human race would operate primarily through the agency of female choice in marriage—"to the cultivated minds and pure instincts of the Woman of the Future" (Wallace, 1890, p. 337). Twenty-three years later Wallace reiterated his view that sexual selection could be an important eugenic force in future human evolution, stating that in the past

> sexual selection tended to act, if at all, prejudically, through polygamy, prostitution, and slavery, though it possesses the potentiality of acting in the future so as to ensure Intellectual and Moral Progress, and thus elevate the race to whatever degree of civilisation and well-being it is capable of reaching in earth-life. (Wallace, 1913, p. 127)

Darwin's theories of natural selection and sexual selection were opposed by many scientists who were neo-Lamarckians. The most extensive argument for the evolution of secondary sexual characters by the inheritance of acquired characters was published by the British physiologist and neo-Lamarckian J. T. Cunningham (1859-1935) in 1900 (excerpts in Paper 23). It is appropriate that this volume, which covers the history of sexual selection theory before the rediscovery of the Mendelian principles of genetics, should conclude with a paper that opposes Darwin's theory of sexual selection primarily because "it does not account for the origin of the variations it assumes" (p. 30 in Paper 23), summarizes and rejects various alternative theories that have been proposed to explain the evolution of secondary sexual characters (Wallace, Paper 20; Stolzmann, 1885; Geddes and Thomson, 1889; Eimer, 1890), and contends that the theory of the inheritance of acquired characters best harmonizes with the then currently known scientific facts (p. 42 in Paper 23). The widespread opposition to Darwin's theory of sexual selection at the beginning of the twentieth century has been summarized by Morgan (1903) and Kellogg (1907). The rediscovery of the Mendelian principles of inheritance in 1900 and the subsequent genetic research that followed led to the scientific collapse of Lamarckian theories of adaptive evolution by the inheritance of acquired characters.

In the *Origin of Species* and *Descent of Man* Charles Darwin contended that "variation distinguishing subspecies and 'allied' (closely related, recently diverged) species very commonly involves variation in sexually selected characters" (West-Eberhard, 1983, p. 156). With the notable exceptions of the writings of Karl Pearson (1900, pp. 418, 421-425) and John Gulick (1905, pp. 83, 200-202), the sexual selection aspect of Charles Darwin's theory of speciation was completely ignored by biologists and played no important role in the formation of modern theories about speciation until well after the construction of the modern evolutionary synthesis (West-Eberhard, 1983). Sexual selection theory was neglected or rejected by biologists until recently for a variety of reasons including the attractiveness of such alternative hypotheses as selection for recognition of the opposite sex, selection for the physiological synchronization of the reproductive systems of potential mates, and selection for reproductive isolation (West-Eberhard, 1983). The ideas of Kropotkin (1903), concerning the relative importance of mutual aid and competition, led many twentieth-century biologists (Wheeler, Allee, etc.) to champion selection for the benefit of species theories of animal behavior and not to take Darwin's theory of sexual selection very seriously.

A companion volume on sexual selection theory and the rise of sociobiology will include papers concerning the widespread rejection of sexual selection theory at the beginning of the twentieth century, the very minor role that was assigned to sexual selection by the neo-Darwinian biologists who constructed the modern evolutionary synthesis during the 1930s and 1940s, and the major role that sexual selection theory has played in the development of sociobiology. Additional information concerning the application of sexual selection theory to human beings will be published in future Benchmark Papers in Systematic and Evolutionary Biology volumes on human evolution by selection and cultural evolution by selection.

The scientific status of evolution by selection theory at the beginning of the twentieth century has been reviewed by Provine (1971) and Bajema (1982, 1983). A forthcoming volume on the theory and measurement of natural selection in the twentieth century will reprint papers that have contributed to the development of natural selection theory since 1900.

Charles Darwin's last defense of his sexual selection theory (Paper 20) was read at a scientific meeting on April 19, 1882, the day he died. The cultural meaning of natural selection has evolved during the century following Darwin's death to encompass all of the ecological interactions that produce differential reproduction of genes regardless of whether the genetic change is brought about by differences in mortality during the prereproductive phase of life or by differences in fertility among those individuals who survive to reproductive age (Mayr, 1982, p. 596). Consequently, sexual selection has come to be viewed by most biologists as a special category of natural selection. This has led Otte (1979, p. 15) to conclude that "the tendency to view sexual selection as something distinct from natural selection is basically without merit" and that "Since the various components do not operate independently of one another, any attempts to rigidly categorize selective agents or to raise one component to the same rank as natural selection is bound to be confusing." Nonetheless Darwin made a wise decision when he coined the metaphor "sexual selection" to draw attention to those ecological interactions an organism has with other members of its own species, which generate selection favoring adaptations beneficial to the individual but harmful to the species. That Darwin's decision was the right one is attested to by the long history of opposition to sexual selection theory, which has only recently been broken by an explosion of scientific interest in sexual selection as a theory to both explain and predict behavior. An adequate scientific understanding of the role sexual selection has and is playing in favoring human behaviors that are beneficial to certain individuals or groups while harmful to many other human beings is crucial if we are to be successful in our attempts to reduce the adverse effects of human conflict both within and between nations.

REFERENCES

Bajema, C. J., ed., 1976, *Eugenics: Then and Now,* Benchmark Papers in Genetics, vol. 5, Dowden, Hutchinson & Ross, Stroudsburg, Pa., 400p.

Bajema, C. J., ed., 1982, *Artificial Selection and the Development of Evolutionary Theory,* Benchmark Papers in Systematic and Evolutionary Biology, vol. 4, Hutchinson Ross, Stroudsburg, Pa., 361p.

Bajema, C. J., ed., 1983, *Natural Selection Theory: From the Speculations of the Greeks to the Quantitative Measurements of the Biometricians,* Benchmark Papers in Systematic and Evolutionary Biology, vol. 5, Hutchinson Ross, Stroudsburg, Pa., 384p.

Broca, P., 1872, Revue Critique: "Les Selections, La Descendance de l'homme et la Selection sexuelle" par Darwin et *"La Selection* naturelle, Essais" par A. Wallace, *Rev. Anthropol.* **1:**683–710.

Cunningham, J. T., 1900, *Sexual Dimorphism in the Animal Kingdom: A Theory of the Evolution of Secondary Sexual Characters,* A. & C. Black, London, 317p.

Darwin, C. R., 1871, 1874, *The Descent of Man and Selection in Relation to Sex,* J. Murray, London.

Geddes, P., and J. Thomson, 1889, *The Evolution of Sex,* W. Scott, London, 322p.

Groos, K., 1898, *The Play of Animals,* Appleton, New York, 341p.

Gulick, J., 1905, Evolution: Racial and Habitudinal, *Carnegie Inst. Washington Publ. 25,* 269p.

Haeckel, E., 1866, *Generelle Morphologie der Organismen,* 2 vols., G. Reimer, Berlin, 606p., 602p.

Haeckel, E., 1868, *Naturliche Schopfungsgeschichte (The History of Creation,* 6th English ed. Appleton, New York, 1914, 422p., 544p.).

Haekel, E., 1874, *Anthropogenie, oder Entwickelungsgeschichte des Menschen,* Engelman, Leipzig (*The Evolution of Man,* Appleton, New York, 1904, 467p., 504p.).

Jackson, R. R., 1980, The Mating Strategy of *Phidippus johnsoni* (Araneae, Salticidae): II. Sperm Competition and the Function of Copulation, *J. Arachnol.* **8:**217–240.

Jackson, R. R., 1981, Relationship Between Reproductive Security and Intersexual Selection in a Jumping Spider *Phidippus johnsoni* (Araneae: Salticidae). *Evolution* **35:**601–604.

Kellogg, V. L., 1907, *Darwinism Today,* Henry Holt, New York, 403p.

Kottler, M., 1980, Darwin, Wallace, and the Origin of Sexual Dimorphism, *Am. Philos. Soc. Proc.* **124:**203–226.

Kropotkin, P., 1902, *Mutual Aid: A Factor of Evolution,* McClure, Phillips and Co., New York, 348p.

Marchant, J., ed., 1916, *Alfred Russel Wallace: Letters and Reminiscences,* Harper, New York, 507p.

Mayr, E., 1982, *The Growth of Biological Thought,* Harvard University Press, Cambridge, Mass., 974p.

Montgomery, T. H., 1910, The Significance of the Courtship and Secondary Sexual Characters of Araneads, *Am. Nat.* **44:**151–177.

Morgan, C. L., 1890, *Animal Life and Intelligence,* Arnold, London, 512p.

Morgan, C. L., 1896, *Habitat and Instinct,* Arnold, London, 351p.

Morgan, C. L., 1900, *Animal Behaviour,* Arnold, London.

Morgan, T. H., 1903, *Evolution and Adaptation,* Macmillan, New York, 470p.

Nutting, G., 1891, Some of the Causes and Results of Polygamy Among the Pinnipedia, *Am. Nat.* **25:**103–112.

Otte, D., 1979, Historical Development of Sexual Selection Theory, in *Sexual Selection and Reproductive Competition in Insects,* M. Blum and N. Blum, eds., Academic Press, New York, pp. 1–18.

Pearson, K., 1896, Contributions to the Mathematical Theory of Evolution. Note on Reproductive Selection, *R. Soc. London Proc.* **65:**301–305.

Pearson, K., 1900, *The Grammar of Science,* 2nd ed., Black, London, 548p.

Peckham, G. W., and E. G. Peckham, 1890, Additional Observations on Sexual Selection in Spiders of the Family Attidae, with Some Remarks on Mr. Wallace's Theory of Sexual Ornamentation, *Wisconsin Nat. Hist. Soc. Occas. Pap.* **1:**117–151.

Peckham, G. W., and E. G. Peckham, 1909, Revision of the Attidae of North America, *Wisconsin Acad. Sci., Arts Lett. Trans.* **16**(5):355–646.

Peckham, M., ed., 1959, *The Origin of Species by Charles Darwin, A Variorum Text,* University of Pennsylvania Press, Philadelphia, Pa., 816p.

Poulton, E. B., 1890, *The Colours of Animals: Their Meaning and Use Especially Considered in the Case of Insects,* Kegan Paul, Trench, Trubner, London, 360p.

Poulton, E. B., 1896, *Charles Darwin and the Theory of Natural Selection,* Cassell, London, 224p.

Rhoads, S. N., 1890, Probable Causes of Poygamy among Birds, *Am. Nat.* **24:**1024–1036.

Romanes, G. J., 1892, 1901, *Darwin, and After Darwin, Vol. I, The Darwinian Theory,* 3rd ed., Open Court, Chicago, 460p.

Sicard, H., 1892, *L'Evolution Sexuelle dans L'Espece Sexuelle,* J. B. Bailliere, 318p.

Stolzmann, J., 1885, Quelques remarques sur le Dimorphisme Sexuel, *Zool. Soc. (London) Proc.* **28:**421–432.

Wallace, A. R., 1871, Review of "The Descent of Man and Selection in Relation to Sex," *The London Academy,* March 15, pp. 177–183.

Wallace, A. R., 1878, *Tropical Nature and Other Essays,* Macmillan, London, 356p.

Wallace, A. R., 1913, *Social Environment and Moral Progress,* Cassell, London, 164p.

West-Eberhard, M., 1983, Sexual Selection, Social Competition and Speciation, *Quart. Rev. Biol.* **58:**155–183.

Westermarck, E., 1891, *The History of Human Marriage,* Macmillan, London, 644p.

21

Reprinted from pages 3–60 of *Observations on Sexual Selection in Spiders of the Family Attidae,* Natural History Society of Wisconsin, Milwaukee, 1889

OBSERVATIONS ON SEXUAL SELECTION IN SPIDERS OF THE FAMILY ATTIDÆ.

GEORGE W. AND ELIZABETH G. PECKHAM.

Introduction.

Mr. Wallace, in his well known essay on *Colors of Animals,* remarks that color *per se* may be considered normal and needs no special accounting for; that amid the constant variations of animals and plants color is ever tending to vary and to appear where it is absent; and that natural selection is constantly eliminating such tints as are injurious to the species while it preserves and intensifies such as are useful; and in opposition to Darwin he has argued that the sexual diversity of color, common in many animals, has its primary cause in the special need of protection for the female, which represses in her the bright colors that are normally produced in both sexes by general laws. Or, to put it in another way, he starts with the fact of the variability of color in animals of both sexes and says that in the female, where greater protection is needed, the color is toned down or eliminated, while in the male, the need for protection being less, the color may be preserved and intensified. Mr. Wallace has supplemented this theory by another factor; he now holds that the frequent superiority, to use his own words, of the male bird or insect in brightness or intensity of color, even where the general coloration is the same in both sexes, is primarily "due to the greater vigor and activity and the higher vitality of the male. * * * This intensity of coloration becomes most developed in the male during the breeding season when the vitality is at a maximum. * * * The greater intensity of colors in the male, which may be termed the normal sexual difference, would be further developed by the combats of the males for the possession of the

females. The most vigorous and energetic usually being able
to rear most offspring, intensity of color, if dependent on or correlated with vigor, would tend to increase. But as differences
of color depend upon minute chemical or structural differences
in the organism, increasing vigor acting unequally on different
portions of the integument, and often producing at the same
time abnormal developments of hair, horns, scales, feathers,
etc., would almost necessarily lead also to variable distribution
of color and thus to the production of new tints and markings.
These acquired colors would *   *   * be transmitted to both
sexes, or to one only, according as they first appeared at an
early age, or in adults of one sex; * * * but in all cases where
an increasing development of color became disadvantageous to
the female, it would be checked by natural selection; and thus
produce the numerous instances of protective coloring in the
female only, which occur in these two groups, birds and butterflies."[1]

We have here two theories offered to explain sexual differences in color: the first is, that natural selection modifies
color in the female for purposes of protection; and the second,
that color may be produced or intensified where there is a surplus of vital energy, as in male animals generally, and sometimes in the females, and more especially at the breeding
season. We will here consider the second theory, since Mr
Wallace regards this as the more important in making intelligible cases of more brilliant coloring in the male as compared
with the female.

What is meant by an excess of vital energy is not quite
clear. Does this term imply that the colored modifications of
the integument represent the excess of nutriment over expenditure? This seems scarcely probable, and yet what other
interpretation is to be put upon such a statement? Supposing
this interpretation to be correct, if the color or development of
plumage represents the surplus over ordinary expenditure,
should not the least active animals, rather than the most vigor-

---

1 *Tropical Nature*, pp. 187 and 193-196.

ous, have the greater surplus and consequently the richer ornamentation?

Grant Allen, in his *Colour Sense,* remarks, concerning the ornamental appendages of animals: "Whatever we may think of their functions, we must agree that they are, on the whole, products of a high vitality. They represent part of the excess of nutriment over expenditure. But these dermal adjuncts do not probably take away anything from the effective energies of the organism."[1] He evidently understands Mr. Wallace to mean that these color adjuncts are by-products, or waste, from the other tissues. In this connection he quotes from Mr. Lowne to the effect that "the dermal appendages of reptiles and the feathers of birds, rich in pigment and nitrogen, are probably entirely excrementitious to the other tissues, and, without doubt, depend in great part for their origin on the solid nature of the excretion of the kidneys. Birds especially, leading a very active life, excrete material rich in nitrogen; and the feathers, which are shed periodically, enable them to throw off that element without overtaxing their renal organs."[2]

"Hence," says Mr. Allen, "we can understand why the more active and energetic sex should possess a greater number of highly developed dermal adjuncts, and should often display much brighter colors than the females." This, however interesting it may be as a speculation, has, so far as we are aware, no direct evidence to support it; and knowing so little as we do at present of the functions of the kidneys in birds, and of

---

1 P. 188.

2 The fact that closely related species annually undergo a double moult and others only a single one, and that even in the same species the sexes sometimes differ in their moulting habits, renders this proposition improbable. The habits being usually identical, why should one species depend upon moulting for the disposal of its surplus nitrogen, while in another the burden is borne by the excretory organs alone? Is there a single anatomical fact to countenance such a supposition? Darwin says that there is reason to believe that with certain bustards and rail-like birds, which properly undergo a single moult, some of the older males retain their nuptial plumage throughout the year. In the birds of paradise some have a single moult, some a double, and others, after the moult of the first year, do not cast their feathers again. For other facts bearing on this habit, see *Descent of Man,* Am. Ed., pp. 391–394.

the nature of the pigment in their feathers, it would be prema-
ture to discuss it in this connection.

Let us see how far the hypothesis that brilliant coloring is
correlated with high vitality is supported by facts. Wallace
makes the activity and pugnacity of an animal the criterion of
its vitality ; and where the male bird takes charge of the eggs and
incubates them, he considers this change of habit, along with
the pugnacity of the female, a proof that in such cases she possesses the higher vital energy, pugnacity being the important
factor. "Of the mode of action of the *general principles of
color-development among animals*,"[1] he says, "we have an
excellent example in humming-birds. * * * The more
vivid colors, and more developed plumage of the males, I am
now inclined to think may be wholly due to their greater vital
energy, and to those general laws which lead to such superior
developments even in domestic breeds ; but in some cases the
need of protection by the female while incubating, to which I
formerly imputed the whole phenomenon, may have suppressed a portion of the ornament which she would otherwise
have attained." In view of the importance of this point the
following evidence, offered in its support, seems rather meagre.
"The extreme pugnacity of humming-birds has been noticed
by all observers, and it seems to be to some extent proportioned
to the degree of colour and ornamentation in the species. Thus
Mr. Salvin observes of *Eugenes fulgens*, that it is 'a most pugnacious bird,' and that 'hardly any species shows itself more
brilliantly on the wing.' Again, of *Campylopterus hemileu-
curus*,—'the pugnacity of this species is remarkable. It is very
seldom that two males meet without an aerial battle,'—and 'the
large and showy tail of this humming-bird makes it one of the
most conspicuous on the wing.' Again, the elegant frill-necked
Lophornis ornatus 'is very pugnacious, erecting its crest, throwing out its whiskers and attacking every humming-bird that
may pass within its range of vision ;' and of another species *L.
magnificus*, it is said that 'it is so bold that the sight of man

---

1 The italics are ours.

creates no alarm.' The beautifully coloured *Thaumastura Cora* 'rarely permits any other humming-bird to remain in its neighborhood, but wages a continual and terrible war upon them.' The magnificent bar-tail, *Cometes sparganurus*, one of the most imposing of all the humming-birds, is extremely fierce and pugnacious, 'the males chasing each other through the air with surprising perseverance and acrimony.' These are all the species I find noticed as being especially pugnacious, and every one of them is exceptionally colored or ornamented; while not one of the small, plain, and less ornamental species are so described, although many of them are common and well observed species."[1]

Here we have *six species* of humming-birds, given as all that are noticed as especially pugnacious, to establish the wide generalization that there is a causal relation between high vital activity, as shown by fierceness and pugnacity, and brilliancy of coloring, in the family *Trochilidæ*, containing 118 genera and 390 species, of which 340 are brightly colored.[2]

The large family of pigeons gives evidence that makes strongly against the theory. Many of them are conspicuously colored—indeed Mr. Wallace remarks, "in the Malay Archipelago and Pacific islands, they occur in such profusion and present such singular forms and brilliant colors that they are sure to attract attention. Here we find the extensive group of fruit-pigeons, which, in their general green colors, adorned with patches and bands of purple, white, blue, or orange, almost rival the parrot tribe; while the golden-green Nicobar pigeon, the great crowned pigeons of New Guinea as large as turkeys, and the golden-yellow fruit-dove of the Fijis, can hardly be surpassed for beauty."[3] If the high vitality of the humming-birds will

---

1 *Tropical Nature*, pp. 213, 214.

2 Baird, Brewer and Ridgway, *N. A. Birds*, say that about fifty species are plainly colored.

3 *Loc. cit.* p. 103. All the pigeons build open nests and the males take part in incubation. In the case of the humming-birds, which also build open nests, Wallace has abandoned, in part, the factor of protection to the female during incubation, since, in a number of the most beautiful species, the sexes are alike, and, as Darwin says, "in the majority the females, though less brilliant than the males, are brightly

explain the unusual development of color in both sexes, and will sometimes override the action of natural selection in keeping down the brightness of the female, this is evidently not the case where the pigeons are concerned, since they are not remarkable for activity nor pugnacity, and are notoriously liable to destruction by many enemies. The further fact that the male is more highly ornamented than the female, and yet assists in incubation, is still more out of harmony with the hypothesis. That the presence of ornaments or gaudy tints is not necessarily correlated with high vitality in birds is shown by the Barbets, which are "rather clumsy, fruit-eating birds," and are clothed in green, diversified by the most vivid patches of yellows, reds and blues.[1]

Again, in the birds of paradise there seems to be no relation between pugnacity and color. Mr. Wallace, in speaking of the splendid Great Bird of Paradise, which he studied in the Aru Islands, says that they congregate at " *sácaleli* or dancing parties," held in certain trees in the forest. " On one of these trees a dozen or twenty full-plumaged male birds assemble together, raise up their wings, stretch out their necks, and elevate their exquisite plumes, keeping them in a continual vibration. Between whiles they fly across from branch to branch in great excitement, so that the whole tree is filled with waving plumes in every variety of attitude and motion."[2] Although these birds were carefully observed not a word is said of their fighting nor of any display of pugnacity. In regard to the Red Bird of Paradise, he mentions having kept a number of the magnificent male birds in the same cage; this he could not have done had they quarreled to any extent.[3] He also mentions " the large cage " of two specimens of the Les.

---

colored." Does not the same reasoning hold good with the pigeons and a host of other birds where the nest is open and the female conspicuous? A fair consideration of the facts seems to us to confirm Darwin's supposition that the habit, common with bright colored birds, of using covered nests was acquired after, rather than before the development of the color.

1 *Tropical Nature*, p. 105.
2 *Malay Archipelago*, p. 446.
3 *Loc. cit.*, p. 536.

ser Bird of Paradise, which he took with him to England, as
though he kept them together.[1] If the trait of pugnacity is
so closely related to brilliant ornamentation in the humming-
birds, why should it not not be so in the birds of paradise?
And why do we find brilliant color in many birds that have no
more—indeed rather less—vigor than the soberly dressed birds
of prey? Since, of all living naturalists, Mr. Wallace is undoubtedly the most competent to discuss this theory of the
correlation of great vitality and bright color, it is surprising
that he can find so little evidence in its favor.

Perhaps the most difficult fact to reconcile with the theory
is the absence of ornamentation and bright color in the bats.
They have wide expanse of integument, and great activity,
the conditions specified by Mr. Wallace for the development of
gaudy pigment, and nothing, apparently, in their habits to keep
it down; but, except in the frugivorous bats, we find little difference between the sexes, nor is there any appreciable approach to bright colors. As Darwin remarks, it deserves attention as bearing on the question whether bright colors are serviceable to male animals from being ornamental, that only in
frugivorous bats is the sense of sight well developed, and it is
only in this group that we find any color.[2]

In the Araneides the great number of species and the wide
differences between the several groups in habits and in amount
of ornamentation, offer unusual facilities for testing Wallace's
two theories of sexual color.

There is a common impression that among spiders
there is little development of ornamentation, and that
they are, as a rule, inconspicuous and dull. This is far from
being true. Wallace, in his *Tropical Nature*, says: "The
small jumping spiders are also noticeable from their immense numbers, variety and beauty. They frequent foliage and
flowers, running about actively in pursuit of small insects; and
many of them are so exquisitely colored as to resemble jewels

---

1 *Loc. cit.*, p. 557.
2 *Descent of Man*, p. 534.

rather than spiders."[1] In the Malay Archipelago, also, he was
struck by the "abundance and variety of the little jumping
spiders, which abound on flowers and foliage, and are often perfect gems of beauty."[2] Bates, too, in *The Naturalist on the Ama-
zon*, says that "the number of spiders ornamented with showy
colors was somewhat remarkable."[3] A large collection of spiders from the tropics is sure to contain as great a proportion of
beautifully colored specimens as would be found among an
equal number of birds from the same region. Let us, then, endeavor to apply to them the hypothesis, that the brighter color
of the male is due to his greater activity and vital force.

Beginning with the most brilliant family, the *Attidæ*, we find
that the females are, with few exceptions, larger, stronger and
much more pugnacious than the males. Some four years ago
we placed two females of *Phidippus morsitans* together in a glass
jar. No sooner did they observe each other than both prepared
for battle. Eyeing each other with a firm glance they slowly
approached, and in a moment were locked in deadly combat.
Within a few seconds the cephalothorax of one was pierced by
the fang of the other, and with a convulsive tremor it relaxed
its hold and fell dead. We placed together, in all, four females,
and in each instance the fight was short but even to the
death. Subsequently, we put in a well-developed male,
which, though smaller, was compactly built and apparently
strong enough to bring the virago to terms; but, to our surprise, he seemed alarmed and retreated, trying to avoid her;
she, however, followed him up, and finally killed him. We
have observed the same habits in *Phidippus rufus*. In *Dendryphantes elegans* the female is nearly a third larger than the
male. During the past summer we kept a number of this
species, males and females, together in a large mating-box, and
were much struck by the greater quarrelsomeness of the
females; they would frequently go out of their way to chase
each other, and they were much more circumspect in approach-

---

1 P. 97.
2 *Malay Archipelago*, p. 437.
3 P. 54.

ing each other than were the males. In *Icius mitratus* neither
sex was especially pugnacious, but the male was as little so as
the female. In *Synageles picata* the females never came near
each other without some display of hostility, though they did
not actually fight.

In several species of *Xysticus*—as *ferox* and *gulosus*—the
females are savage and ready to attack anything that comes
in their way, while the males are smaller and more peaceable.
De Geer tells of a male spider that, "in the midst of his preparatory caresses was seized by the object of his attentions, enveloped by her in a web, and then devoured, a sight which, as he
adds, filled him with horror and indignation."[1]   The Rev. O.
P. Cambridge holds that the greater ferocity on the part of the
female in the genus *Nephila* has led, through the action of
natural selection, to the extreme reduction in the size of the
males.[2] In each of two species of *Lycosa*, whose mating
habits we were endeavoring to discover, two males were
destroyed by a single female.

Not all female spiders are savage and quarrelsome. In
some genera, as *Linyphia*, the two sexes live happily together
in the same web; but Hentz, after twenty years' study of North
American spiders, says that "there is less ferocity in the spiders of this division than in any other of the family. It is the
only sub-genus in which the male and female may be seen har-
moniously dwelling together."[3]   Although subsequent investigation has made it necessary to qualify this statement, the

---

1 Kirby and Spence, *Entomology*, Vol. I, p. 280, 1818.

2 *Proc. Zool. Soc.* 1871, p. 621.   Simon has the following interesting remarks on
sexual differences in size in Epeiridæ: "Dans les genres où l'inégalité est faible, le
nombre des mâles paraît égal à celui des femelles, car à l'époque de l'amour ces *Epeiridæ* se rencontrent régulièrement par paires; mais, dans les genres où il y a grande
disproportion, le nombre des mâles est beaucoup plus considérable, car il n'est pas
rare de voir quatre ou cinq individus de ce sexe courtiser une seule femelle. Ces petits
mâles sont adultes les premiers, mais la durée de leur vie paraît très-courte, car après
l'époque de la reproduction ils disparaissent complètement; ils ne construisent point
de toile propre; mais ils se tiennent à proximité des endroits habités par la femelle,
attendant le moment propice pour l'accouplement, qui a lieu au milieu de la toile de
celle-ci est qui est toujours précédé de longues hésitations. *Les Arachnides de France*,
I, p. 20.

3 *Spiders of the United States*, p. 132.

broad fact remains that, as a rule, the females are more powerful and more pugnacious than the males. Walckenaer, Menge,
Hentz and others give numerous instances where the male
meets his death through the fierceness of his mate; in fact, the
danger is so imminent that after mating it is the habit in several genera (*Epeira* and *Tegnaria* are mentioned by Walckenaer)
for the male to retire with precipitation from the web of the
female, as a reasonable precaution. The relations between the
sexes have been admirably characterized by Romanes in *Animal Intelligence*, where he says: "In many species the male spider in conducting his courtship has to incur an amount of personal danger at the hands (and jaws) of his terrific spouse
which might well daunt the courage of a Leander. Ridicu-
lously small and weak in build, the males of these species can
only conduct the rites of marriage with their enormous and
voracious brides by a process of active manœuvering, which, if
unsuccessful, is certain to cost them their lives. * * * There
is no other case in the animal kingdom where courtship is
attended with any approach to the gravity of danger that is
here observable."[1]

It might be supposed that in spiders the usual conditions
are reversed and that, as in some birds, the females are more
beautifully colored as well as more pugnacious than the males.
This, however, is not the case. Even where the coloration of
the two sexes is similar the tints of the male are usually
brighter; and in many cases, especially among the *Attidæ*, the
female is dull-colored, while the weak and unaggressive male is
extremely brilliant.

There is a family of spiders, the *Gasteracanthidæ*, compris-
ing a number of genera and several hundred species, widely
distributed and very rich in individuals. In the whole order
of spiders there is no group where the females are so universally remarkable for inactivity and sluggishness of movement.
After the web is made she remains, nearly all the time, standing motionless in the center. The males, so far as they are

---

1 P. 204.

known, are smaller and much more active in their habits. Contrary to what we should expect on the theory, however, the males are, as compared with the females, inconspicuous and plainly colored; the females being almost always strikingly and often very brilliantly adorned with black, yellow, red, blue and white. If it be objected that the *Gastera-canthidæ* do not entirely meet the point in question, since their colors have been developed as a warning of their inedibility, there are scores of edible species among the web-building spiders which are brilliantly colored and, at the same time, are very inactive; as in the genera *Epeira, Meta, Tetragnatha, Theridium, Linyphia* and others.

Turning to the family *Lycosidæ*, with its numerous species, just the reverse appears, since it is among these vagabond or wolf-spiders that we find the most vigorous species, ever running about and full of restless activity. We know a good deal of their habits and of the courage of the females in defending their young; but, contrary to what we should expect, this family presents very little color development, dull grays, browns and blacks prevailing. The *Agelenidæ* are very active in their movements, throwing themselves upon their prey with great vigor, and depending upon their strength, and not upon web-lines, to hold it, even when it is large and powerful. These spiders are also dull colored.

The spiders, then, seem well adapted to disprove the proposition that there is a causal relation between vital activity and color development, since in the sedentary groups, while many of the species are plainly colored, there are nearly as many that present the most beautiful tints; and, on the other hand, some of the wandering and very active groups are, for the most part, clothed in sombre hues.

Turning to the other question, how far have the females had their color toned down, as compared with the males, for purposes of protection during the nesting period? If we examine the *Attidæ*, where there are marked differences in the coloration of the sexes, we find that all the females remain under a thick, web-like covering, with their cocoons, until the eggs are hatched—that is to say, all the species of this family have covered nests. Many of these covered nests are occupied by dull-colored females, the males of the same species being showily attired. The same is true, to a great extent, among the *Thomisidæ*, where the females, often protectively colored, remain near the eggs in a covered nest. The protectively colored *Lycosidæ*, it is true, have practically open nests, as they usually carry the eggs about with them. In other families the habits vary, some leaving the eggs to their fate as soon as the cocoon is formed, while others give it much or little attention, as the case may be. A general survey of the facts shows no relation between color development in the females and their nidifying habits, and it is highly probable that in spiders, as in birds, the color was developed prior to the formation of these habits.

We must, then, seek further for an explanation of sexual coloring in spiders, since here we have present to account for it neither special vitality on the part of the male, nor need of protective coloring, while nidifying, in the female.

MOULTING HABITS.

There is no group of facts that brings out the remarkable similarity between birds and spiders in color development more prominently than that which is gained from a study of the moulting habits of the two classes. In spiders, as in birds, the young very often differ from the adults, and in many species where the sexes differ when adult, the male being the brighter, they are alike until they reach maturity, when the male, along with his sexual development, acquires his brilliant color. Cuvier formulated, under several rules, the various changes that the plumage of birds undergoes from the nestling to the adult. These rules were extended by Blyth; and Darwin, in *The Descent of Man*, not only amplified and added to them, but also submitted them to a thorough analysis and discussion in order to discover the causes of the phenomena. In looking over systematic works we unfortunately obtain little or no information about the moulting habits of spiders. Most of the species have been described by workers living in the large cities of Europe, from collections coming to them from distant parts of the world, so that an observation of habits was impossible. Then, too, such collections contain only adult or nearly adult forms, since both systematists and collectors have thus far given their attention only to these. There are, therefore, accessible on this subject, little more than the impressions of several high authorities in arachnology as published by Darwin. For some years past we have been accumulating data as to the differences in form and color between the young and the adults, and also between the two sexes in the adult stage in the same species.

Before giving the facts that we have thus far obtained it may be well to remind our readers that spiders, shortly after hatching, cast the skin, and that this moulting of the integument, including the outer coat of the eyes, is repeated, the number of times varying in the different species and possibly in the two sexes. It is probable that the *Attidæ* moult from seven to eleven times before reaching maturity. In *Dendryphantes capitatus* we have counted ten moults and the spider was still immature. If one examines any of our spiders soon after they are hatched he will verify Dr. McCook's generalization that " the color of young spiders is almost without exception bright yellow or green, whitish or livid."[1] Soon after this stage, probably at the third or fourth moult, colors appear, distributed in patterns characteristic of the species, and as the spiders continue to advance in age and make their successive moults, still other and more marked changes may be noted.

Let us now pass to the consideration of the classes of cases under which the differences and resemblances, in color and form, between the young spiders and the adults of one or both sexes may be arranged. It is true among spiders as among birds, that the several classes pass into each other; and that

---

[1] Proc. Acad. of Nat. Sci. of Phila., 1888, p. 172.

when the young resemble their parents, the resemblance, although very strong, is not so complete as to render them exactly alike.

CLASSES OF CASES.

I. When the adult male is more conspicuous than the adult female the young of both sexes in color and form closely resemble the adult female.

II. When the adult female is more conspicuous than the adult male the young of both sexes, in color and form, more closely resemble the adult male than they do the adult female, especially in the earlier moults.

III. When the adult male resembles the adult female the young of both sexes resemble the adults.[1]

In Darwin's discussion of the subject of sexual selection, he considered in great detail how far the moulting habits of birds tended to support his theory that the differences between the two sexes are attributable to female selection, rather than to natural selection acting upon the greater need for protection on the part of the female. In his profound discussion of the laws of heredity he formulates two general propositions. *First:* That variations appearing early in the life of an organism would tend to be transmitted to the offspring of both sexes. *Second:* That variations appearing late in life would be limited to the sex in which they first appeared, and would tend to appear at a corresponding age; the exception being that they might appear at an earlier age in the offspring than they did in the first instance. When the great complexity of the subject is considered, and the way in which natural selection must have sometimes modified sexual selection is taken into account, it is remarkable how fully the moulting habits of birds confirm his generalizations; and it is of the highest interest to inquire how far the moulting habits of spiders are also consistent with them.

Class I includes the cases where the adult male spider is

---

1 These classes are slightly modified from Darwin, *Descent of Man,* p. 466.

more conspicuous than the adult female, the young of both sexes resembling the adult female, both in color and form—or, at least, resembling her much more than they do the adult male. A good example of this class is *Phidippus johnsonii*, where the female has the abdomen red and black, with a white base and some white dots, while that of the male is bright vermillion red with sometimes a white band at the base. The young of both sexes resemble the mother until the last moult, when the males assume their bright livery. *Philæus militaris*, a very common *Attus*, is another illustration; in the male the cephalothorax and abdomen are bright bronze brown, the former with a wide, pure white band on each upper side and a white spot on the center of the head, the latter with a wide, white band around the base and sides; the female has a brown body covered over with white and gray hairs, which form a more or less distinct pattern of lines and spots. To give a better idea of this difference, let us suppose a male bird with the body, neck and head bright bronze brown, and the wings and a patch on the head pure white, with a female having mottled white and brown plumage.

Hentz described the female of his *Plexippus puerperus* as a different species under the specific name *sylvanus*, so little do the sexes resemble each other. *Dendryphantes capitatus* is another species in which there are great sexual differences. As in the last instance, the male and female were described by Hentz as different species. We may suppose that the sexual peculiarities of the male have been only recently acquired in *capitatus*, since he sometimes retains the markings and color of the female, these being proper to him in the immature stage. In *Icius palmarum* the sexes are very different. We shall, later on, describe the differences in the face and falces, so that here it is enough to say, that in the male the whole body is bronze brown, covered with short, golden down, while in the female the color is rufus, with black and white markings.

For *Habrocestum splendens* the colored figures of the moults and the adult forms (Plate I) bring out the fact that while the young are not exactly like the adult female they resemble her much more closely than they do the adult male. This is one of our most beautiful males. The highly iridescent scales, which cover the entire body, make it impossible to give, in a painting, a correct idea of its brilliancy, since the color changes in every light. The male only gets this gorgeous livery at the last moult, just as he becomes mature, though in some species the nuptial robe is acquired one moult before maturity.

The family *Attidæ*, from which these illustrations have been taken, is by common consent, placed at the head of the order, and contains among its 1,500 species the greatest amount of sexual difference and the highest development of ornamentation; indeed, as we have seen, Wallace speaks of them as resembling jewels rather than spiders, and Walckenaer says that their species are well marked by the rich diversity of their colors and the variety of the designs which ornament their abdomens.[1]

In the seventy-eight species described in our work on the *Attidæ* of North America we have both male and female in only forty; or, to be more accurate, we have felt warranted in placing males and females together in only forty instances. Doubtless in many cases we have separated the two sexes, making two species out of one, but the difference in color is so great that, without knowledge of their habits, no other course was possible.[2] Of the forty species, we know the moults of thirty_ two, and of these, nineteen species form a group characterized by marked sexual differences, the males being very generally conspicuously colored as compared with the females, while in others only the falces are different. Since the nineteen species represent twelve important genera[3] it would seem that so far as the North American *Attidæ* are concerned, the generalization is

---

1 *Hist. Nat. des Insectes Aptères,* I, p. 481.

2 We have, in the genus *Phidippus,* sixteen species, four pairs, four single males and eight single females. *P. cardinalis,* male, is of a splendid, uniform, cardinal red; to which of the eight plainly attired females he belongs we do not know; and we have the same difficulty in other cases.

3 *Phidippus, Philæus, Plexippus, Dendryphantes, Icius, Habrocestum, Astia, Zygoballus, Synageles, Menemerus, Lyssomanes,* and *Epiblemum.*

well established. If we may add to these a large number of
bright-colored, undescribed species from our own Guatemala
collection, in which the young closely resemble the female, the
generalization is materially strenghtened.

In order to estimate to what extent the sexes differ in the
Attidæ, we tabulated the species, in a number of works, giving
the number of species in which both sexes, and also the number in which one only are described. If it may be assumed
that a collector ordinarily takes as many of one sex as of the
other—and our numerous collections from different parts of the
world confirm this supposition[1]—if the sexes were fairly alike
they would be identified as one species and placed together; if,
on the contrary, the sexual differences were great (the habits
being unknown) many species would be founded on a single
sex.

TABLE OF SPECIES DESCRIBED BY DIFFERENT AUTHORS, ACCORDING
TO SEX. TOTAL 930.

AUTHOR.	LOCALITY.	♂ & ♀	♂	♀
Koch and Keyserling.	Australia	48	47	57
Taczanowski	South America	60	27	26
C. Koch	World, except N. Am. and Europe	3	49	46
Cambridge	Various	16	27	16
Lucas	Algiers	8	13	36
Thorell	Burmah and Indian Archipelago	17	62	42
Simon	Various places, not France	8	47	22
Walckenaer	World, except N. Am. and Europe.	3	11	17
Peckham	North America	40	17	21
Simon	France	94	28	11
Vinson	Madagascar, Mauritius, etc.	..	..	8
		297	328	306

The table shows that in a total of 930 species from all parts
of the world the single males just about balance the single

---

[1] That this is not always true in other families is shown by Stoliczka, who says:
"In collecting *Epeiridæ* I was particularly struck with the very great scarcity of male
specimens; for, among about 200 specimens belonging to about thirty species, there
were not more than five or six males." *Indian Arachnida*, Proc. Asiat. Soc., Vol.
XXXVIII, p. 234. This was probably due to his having collected either before or after
the mating season, when the males are always more rare.

females, and that the same is true, with a few exceptions, in the individual collections. Excepting when the spiders are well-known, as in the case of Simon's *Attidæ of France,* a comparison of the description of the two sexes, when they have been put together, would only serve to show a similarity between the males and females, since it would be only in the species where there was little or no sexual difference that they could be placed together. Where they are well-known, as in the *Attidæ of France,* we find by such a comparison that in thirty-nine species the male is plainly unlike the female, being in twenty-six instances much more conspicuous, while in fifty-five the sexes are similar, or, if they differ, the male is no more conspicuous than the female. These facts are given to make it clear that the sexes very commonly differ, the male being brighter than the female. It is probably not too much to say, that in the *Attidæ,* at least two-fifths of all the species have the male more conspicuous than the female.

Menge[1], in referring to the greater brilliancy of the male of *Micromata ornata,* says that it only assumes its bright color as a "bridal adornment," and in this connection makes the statement that in the families *Thomisidæ* and *Salticidæ* the males are generally more beautifully colored than the females. We have, in North America, several *Thomisidæ* that are like Menge's species in the difference in color between the sexes, and also in that the young males are like the female, and only assume their bright color at the last moult. Darwin remarks on the fact that "the female of *Sparassus smaragdulus* is dullish green, whilst the adult male has the abdomen of a fine yellow, with three longitudinal stripes of rich red;" while young this male resembles the female. The obvious conclusion from these facts is that it is the male that has varied, and this, too, late in life, so that his peculiarities, having been limited to one sex, do not appear in the young. We are not embarrassed in this group by any need on the part of the female for plain colors to protect her during incubation, since, without exception, the cocoon

---

1 *Preussische Spinnen,* II, p. 396.

is so placed as to be concealed, or, as in the *Attidæ*, is covered with a thick layer of web, under which the mother spider remains until the young appear, seldom leaving the nest to obtain food. On the whole, the explanation of sexual differences of color in spiders which is most conformable with facts, is that the males have departed from the usual colors of the genus,[1] and that most probably this departure has been brought about by sexual or female selection.

Thus in the *Habrocestum* group the general coloring of the genus is represented by the females of *cristatum* and *auratum*, which are gray, with oblique whitish bands; in *viridipes* there is a tendency to greater concentration of color, and consequently to stronger contrast, the color being blackish with yellowish white bands. In *peregrinum* and *auratum* we find that the males have gone still further in the same direction, the ground color being deep black and the bands pure white. (Plate 1.) In *splendens* the male has made a still greater departure from the typical coloring of the genus, while the female, also departing therefrom, though not in so great a degree, shows the strongly contrasting black and white that is found on the males of *peregrinum* and *auratum*.

Class II.   When the adult female is more conspicuous than the adult male, the young of both sexes, both in color and form, more closely resemble the male than they do the female, especially in the earlier moults.

This class, while it is found both in spiders and birds, differs widely in the two groups, both in the number of instances and in the causes that have produced them.   While in birds the number of cases in which the females are brighter is

---

1 This is the opinion of Blackwall and of Canestrini.   Thorell, in speaking of the genus *Erigone* (of the family *Therididæ*, a genus that contains an immense number of species, says: "The study of the spiders belonging to this interesting genus has hitherto been comparatively neglected, and this neglect is no doubt to be attributed partly to their diminutive size, and partly to the great similarity prevailing among the *females* of the different species. * * * Many a female is sometimes mated with one, and sometimes with another male; * * * the following lists of synonyms must, unless the contrary be directly stated, be considered as applying only to the *males*, which are comparatively easily distinguishable." *Remarks on Synonyms of Europ. Spiders*, p. 97.

inconsiderable, and while even in these cases the female is but little more conspicuous than the male, in spiders there are numerous species in which the female is decked in the most gaudy hues, her body being at the same time protected by strong, sharp spines, while the smaller male is unarmed and comparatively inconspicuous.[1] (Figs. 1 and 2.) The spiders to which we refer belong to the genera *Gasteracantha*, *Acrosoma*, *Phoronocidia* and others, including possibly 250 species distributed over all parts of the world. The differences between the males and females are shown in Plate III. Only a few males are described in this group, but they all agree in the peculiarities mentioned.

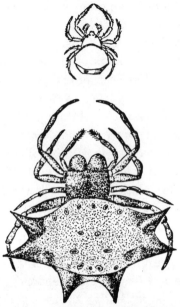

Fig. 1.— Gasteracantha rufospinosa. Upper figure, male enlarged seven times; lower figurer female, enlarged four times (from Marx.)

The perplexing fact in this connection is that while the females at first resemble the males, it is only while they are quite young, since when they are from a quarter to a third grown they begin to assume the adult form and color, and it seems as though characteristics developing so early should have been transmitted to

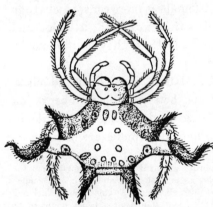

Fig. 2.—Gasteracantha crepidophora, female (from Cambridge.)

---

1 A seeming exception to this rule is G. Cowani, ♂, described in *Proc. Zool. Soc.*, 1882, p. 766. As this case was sufficient to invalidate our generalization we wrote to Mr. Butler, asking him to re-examine the spider in question to make sure that it

both sexes.   After a good deal of consideration we are
inclined to believe that in the early history of the group
the male and female both possessed some such form and
color as is now seen in the adult male, and in the first few
moults of the female; and that afterwards the adult female,
probably on account of some change in habits, varied toward
her present size, form and color under the action of natural
selection.   " As variations occurring late in life," says Darwin,
"and transmitted to one sex alone, have incessantly been taken
advantage of and accumulated through sexual selection in relation to the reproduction of the species; therefore it appears,
at first sight, an unaccountable fact that similar variations have
not frequently been accumulated through natural selection, in
relation to the ordinary habits of life.   If this had occurred
the two sexes would often have been differently modified, for
the sake, for instance, of capturing prey or of escaping from
danger.   Differences of this kind between the two sexes do occasionally occur, especially in the lower classes.   But this implies that the two sexes follow different habits in their struggles
for existence, which is a rare circumstance with the higher
animals."[1]

The other supposition open to us, namely, that the variation from the male form began in young females and were sexually limited from the first, is improbable, in view of the mass
of evidence that variations before maturity are inherited by
both sexes equally.   The habits of the female, standing nearly
all the time exposed in the web, give the clue to the occasion of
her modification.   The habits of the male being different, he
is left unmodified ; he is usually found in less exposed posi-

---

was a male and not a female. Mr. Pocock, who is now in charge of the *Arachnida* in
the British Museum, responded as follows:

BRITISH MUSEUM (Natural History), LONDON, March 23, 1889.

*   *   *   *   *   *   *

You are quite right in supposing that Mr. Butler fell into error in describing
his specimens of this species as being males. They are in reality females. The
distal segments, however, of the palpi are considerably wider than the preceding, and
no doubt the mistake arose from a superficial examination of this appendage.

1 *Descent of Man,* p. 241.

tions, and only for a few days, at the mating time, is to be seen in the web.

Class III. When the adult male resembles the adult female the young of both sexes resemble the adults.

The greater number of species in this class have dull colors, which seem to be protective, as in the *Lycosidæ*, many of the *Drassidæ, Epeiridæ, Thomisidæ* and others. The same problems are presented here that are so ably discussed by Darwin in the case of birds belonging to this class. It seems likely that in the orb-weavers a large part of the courtship is conducted by vibrating the web lines, and that the males depend upon their skill in this direction, and not upon color alone, in attracting the female. The modifications of the first pair of legs, so common in the male spiders of the group, might be explained as useful in producing variations in these vibrations. We have some little evidence on this point, but defer consideration of it until treating of mating habits. In the *Attidæ*, when the male is not strikingly colored, it is more than probable that the choice of the female is determined by his antics and grace of movement. *Saitis pulex* is a good example. While in the greater number of instances in this class the colors are dull, we have many species which are brilliant. Thus, in the genus *Homalattus*, many of the species have metallic greens or lovely iridescent blues or violets, while others are soft brown, mottled with white. We have not enough data to throw any light on the very intricate problems here presented, and can only refer the reader to Darwin's work,[1] where the whole subject is admirably discussed. If, therein, he will substitute *spider* for *bird*, whenever the latter word appears, he will see how full of difficulty is the whole subject.

We have other cases that come under several classes, but as the instances are few, we will simply describe them individually. In *Hasarius hoyi* the adult male is more conspicuous than the adult female, which, indeed, was first described as a separate species under the name *pinus*. The young, very early,

---

1 *Loc. cit.,* p. 481.

differ from each other according to sex, the young males resembling the adult males, and the young females the adult females. In this case the law of inheritance at corresponding ages must be supposed to have failed, the young males inheriting their differential peculiarities at an earlier age than that at which the variation first appeared in their male ancestors.

Phidippus rufus, when mature, is a brick-red spider, the male being considerably brighter than his consort. When about one-seventh grown, and after the third or fourth moult, the young are very dark brown, with light yellow legs. Some moults later, they are reddish, with narrow, oblique whitish bars on the sides of the abdomen, and two dark bands on the dorsum, on each of which is a row of white dots. The appearance of the spider changes but little during the next four moults, but after the last—the tenth—both male and female become mature and acquire the adult color. The meaning of the fifth moult—that with the uniform brown body and yellow legs—we are unable to explain. The appearance of the female after the fifth moult is similar to that of many other females in the genus, and her final change is probably due to a transference to her of the male color, which, judging from the moults, must have appeared late in life.

All the cases under the first and third classes are intelligible if we suppose that the females have selected the more conspicuous males, and that when variations occurred late in life they were limited to the sex in which they first appeared. There are, however, many cases in which it is probable that the color variation of the male has modified the female in a greater or less degree, and this accounts for the instances in which the female, as well as the male, is showily colored, especially since the showy color in these females usually tends to approach the coloring of the males, and to depart, in the same proportion, from the normal coloring of the genus. The second class, in which the females are more brilliant than the males, is the only one in which the moulting habits of spiders are not strikingly similar to those of birds; and it seems in the highest

degree probable that where there is so close a resemblance
between two such complex sets of phenomena, in widely
separate classes, they can only have been brought about by a
common cause.

SECONDARY SEXUAL CHARACTERS.

It is a noticeable feature in the secondary sexual peculiarities of spiders, beside those of color, which have already
been sufficiently dwelt upon, that they are very commonly
found in the falces, clypeus, palpi and first pair of legs—that is
to say, in those parts of the animal that are plainly in view
when the male is paying court to the female; and it is a fact
of great significance that even in the species where sexual differences are reduced to a minimum we usually find a modification of one or more of those parts which serve to render
the male somewhat more conspicuous or showy than the
female. *Synagelcs picata* is a case in point, the sexes being nearly
alike, but the male having the first legs flattened and brilliantly
iridescent.   In several species of the genus *Lyssomanes* the sexes

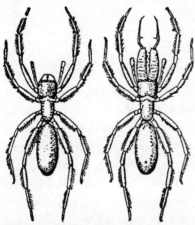

only differ in the length of
the falces and that of the first
pair of legs or palpi.[1]

The ant-like spiders are
notable for differences in the
falces of the two sexes. *Salticus formicarius*, (Fig. 3), a
common European species, is
a good illustration, the female
having short, vertical, reddish black falces, while those
of the male are horizontal,
much enlarged, and copper
green in color. In an ant-like

Fig. 3.—Salticus formicarius (from C.
Koch.) Right-hand figure, male; left-hand
figure, female.

spider from Australia, *Synemosyna lupata*, (Fig. 4, see p. 27), Dr.
Koch describes a very curious development in the falces of the
male; their great length, their teeth, and their branching

1 *Lyssomaues viridis, jemineus* and *amazonicus*.

fangs, reminding one of the great mandibles in certain male *Lucanidæ.* The female is undescribed, but we know, in this genus and in others closely related, so many males that have enlarged falces while those of the female are short, that this, too, is doubtless a sexual characteristic.[1]

Cambridge has described two spiders from Ceylon[2] which are remarkable for the great elongation of the male falces, these being in both species longer than the cephalothorax, while the fangs are toothed and equal in length to the falx. In both species, too, the falces seem to be attractively colored, as in one they are "dark black-brown and shining," and in the other "very slightly and transversely rugose, and shining in some lights with an opaline hue." It is true in many males that the

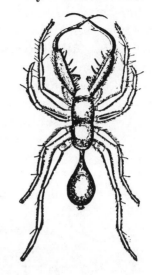

Fig. 4.—Synemosyna lupata. Male (from L. Koch).

falces are not only greatly enlarged but are also brilliantly colored; and in this connection it is of high importance to note that the brightly colored hairs or metallic scales, as well as the protuberances, are either on the anterior surface, or in some way so placed as to be plainly in view from in front.

In a common spider of the Southern and Eastern United States *Icius palmarum* Hentz we have a good illustration. In the male the falces are compressed and horizontal and are three times as long as the face, the fang equaling the falx in length. The front surface of the falces is dark bronzy rufus and on each outer edge is a wide band of snowy white hairs. In

---

1 We have examined 49 species of ant-like spiders, but in no instance have we found a female with long falces, while we know of 19 males in which they are much lengthened. One might have expected to find that an occasional female had inherited the modification from the male, but it seems to be, in this group, strictly limited to one sex.

2 *Ann. and Mag. of Nat. Hist.,* 1869, pp. 68-70.

the female the falces are vertical and only as long as the face, the fang being equally reduced, and the white hairs are absent. The male is rendered still more striking by the long, snowy white hairs which cover his clypeus, while the forehead and a space just below the first row of eyes is covered with bright red hairs. All this ornamentation is lacking in the female, and the contrast between the showy male and his modestly attired mate is very striking. In the little cosmopolitan zebra spider, *Epiblemum scenicum* we find the same difference in the

size of the falces of the two sexes, the male having them four times as long as the face, while in the female they are only one-and-a-half times as long. Dr. Koch, in his magnificent work *Arachniden Australiens*, figures and describes an Attus *Opisthoncus abnormis* (Fig. 5), with curiously formed falces. Their general color is yellowish brown, but the front surface is coppery red, and toward the inner edges

they are of a pretty, iridescent, bronzy green ; but nature, less generous to the female, has given her only some white hairs over her small and unmodified yellowish brown falces. For

Fig. 5.—Opisthoncus abnormis (from L. Koch). Upper figure. face and falces of male; lower figure, face and falces of female.

purposes of offense or defense, however, the female fang is the more effective of the two, and our fine fellow doubtless hopes more from his beauty than his strength.

In the sub-family *Tetragnathinæ* the falces are long in both sexes, but longer and much more ornamented with various processes and bunches of hair in the male than in the female.[1] (Figs. 6 and 7, see p. 29.) The two figures fairly represent this greater development in the one sex than in the other.

Canestrini remarks on the sexual differences in the falces, saying : "Sometimes they are long and strong, with fine teeth,

---

1 This point has also been noted by J. H. Emerton, *New England Epeiridæ*, p. 298. He says of the *Tetragnathinæ*: "The mandibles, especially in the males, are very long, and toothed on the inner edge."

in the male, while in the female they are short, weak and without teeth."[1] He also speaks of the entirely different colors of
the male and female falces of the same species.   Not infrequently there are two forms among the
males of a species, one with long and modified falces, and another having them
short, and more like those of the female.
In *Pensacola signata*, an Attus from Guatemala, we have two such forms.   To quote
from a former paper of our own: " In the
first, which is a little the larger, the falces
are more than twice as long as the face,
slightly retreating, narrow at the base and
extremity, but dilated in the middle when
looked at from in front.   Near the anterior inner edge, in the
middle, is a strong apophysis or spine in each falx, which
reaches nearly to the end of the fang; fang long and slightly
bent.   In the second form the falces are relatively shorter, and

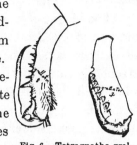

Fig. 6.—Tetragnatha gral-
lator (from Emerton). Left-
hand figure, falx of male;
right-hand figure, falx of
female.

are but very little dilated
in the middle, so that the
curve on the inner edge is
not marked, and the spine
is less than half as long as
in the first form.   We find
others   intermediate   between these two extremes.[3]
Canestrini also refers to the
existence of   t w o   m a l e
forms, and explains them
by the supposition that we

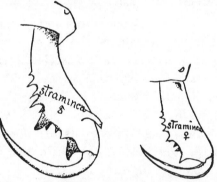

Fig. 7.—Tetragnatha straminea (from Emerton).   Left-hand figure, falx of male; right-hand
figure, falx of female.

here see these secondary sexual characteristics, as it were, in
process of development.[4]

Another curious modification is found in *Icius cornutus*,

---

1 *Caratteri sessuali secondarii degli Arachnidi*, Atti della Soc. Vento—Tren-
tina di Sc. Nat. Padova, I, Fasc. 3, 1873.
2 *Loc. cit.*
3 *On Some New Genera and Species of Attidæ*, G. W. and E. G. Peckham, Proc.
Nat. Hist. Soc. of Wisconsin, 1887, p. 84.
4 This occurs in *Philæus militaris*, *Icius palmarum*, and *Zygoballus bettini.*
Canestrini mentions the genera *Linyphia*, *Theridium* and *Dysdera*.

from Madagascar, where there is long projection from the anterior face of each falx, the two processes extending forward and looking a little like a pair of horns; they are more than half as long as the cephalothorax.[1]

The height of the clypeus[2] was formerly used in the classification of *Attidæ*, but it has more recently been found to be so entirely a sexual peculiarity, as to be of little or no taxonomic value. It is frequently adorned by colored hairs, which fall over the falces, or is diversified by curious patterns formed by bars and patches of down. The very striking appearance that a spider presents with this feature well developed is shown in *Titanattus sævus*, a male from Guatemala. In a

Fig. 8.—Dendryphantes capitatus. Male, face, falces and palpi (from nature).

common North American spider, *Dendryphantes capitatus*, (Fig. 8), the clypeus of the male is conspicuously marked by several white bands; one passes up between the anterior middle eyes from the base of the falces; and two on each side pass back over the cephalothorax. The contrast between these snowy white bands and the dark color of the rest of the face is exceedingly striking. The female has the whole clypeus whitish and is not at all conspicuous. In two species of *Habrocestum, coronatum* and *cæcatum*, the clypeus is covered with bright red hairs. In describing *H. coronatum*, Hentz says that "the bright scarlet spot on its front gives to this spider a whimsical air of fierceness, which is heightened by its attitudes and singular motions."[3] In *Thorellia ensifer*, the male has two bunches of stout dark hairs projecting forward from just above the insertion of the falces, which are not present in the female.[4] *Hyllus pterygodes*[5] has the clypeus, or rather what might be called the cheeks, drawn

---

1 *On Some New Genera and Species of Attidæ from Madagascar*, G. W and E. G. Peckham, Proc. Nat. Hist. Soc. of Wisconsin, p. 31.

2 The *clypeus* is all of the face above the insertion of the falces and below the first row of eyes.

3 *North American Spiders*, p. 65.

4 Koch and Keyserling, *Arachniden Australiens*, p. 1353.

5 *Id ibid.*, p. 1339.

out and gradually coming to a point on each side, so that the face looks, from in front, very much swollen and enlarged. This surface, in both sexes, is covered with bright scales which are somewhat rosy in tint, and the points at the sides are furnished with some stout dark hairs.

In examining the upper part of the face, just above or below the first row of eyes, a number of interesting features may be observed. In one section of the sub-family *Lyssomanæ* most of the species have, for the general color of the body, a tender grass green. In this group the clypeus and the region around the first row of eyes is nearly always adorned with a covering of red hairs, which are sometimes dull, sometimes very bright; this ornamentation is not usually confined to one sex, but in *L. amazonicus* the red is perceptibly brighter in the male, while though in *Asamonea puella* the eye-region of the female shows no red hairs, the male has the forehead covered in the middle by thick, silvery white, and on the sides by reddish hairs, his clypeus, also being unusually high.

The other section of this sub-family often presents dark colors, and here the clypeus and eye-region are more frequently marked with white pubescence or metallic scales than with red hairs; thus in *A. tenuipes* the dark clypeus of the male is covered with highly iridescent scales, and that of the female is light yellow, covered with thick, snowy white hairs.

The bright markings in some of these species have evidently been transmitted to the females through the males. Looking at the group as a whole, it is important to note how frequently the adornment is so placed upon the body as to be brought into view when the spiders face each other.

In *Amycus micans*, (Fig. 9), of which only the male is known, the face is very high, and all its parts are covered with glittering violet, green and golden scales; above the first row of eyes is a transverse band of scale-like hairs, some few longer hairs growing out between; this band is shining, colored

Fig. 9.—Amycus micans. Male, face and falces (from L. Koch).

like the rest of the face.[1]   *Amycus tristriatus* has a high face,
the hairs around the eyes being white in the lower half and
yellowish red above.[2]

In the nine species of the genus *Amycus* described by the
two Kochs, we have only males; in the South American spiders
of this genus described by Simon and Taczanowski,[3] we have
both sexes in four species, and in three of these the clypeus of
the female is low; in the fourth it is said to be " rather high."
The great height of the clypeus in *Amycus* is, in fact, a sexual
peculiarity, although in some instances
it may have been transmitted to the
female.

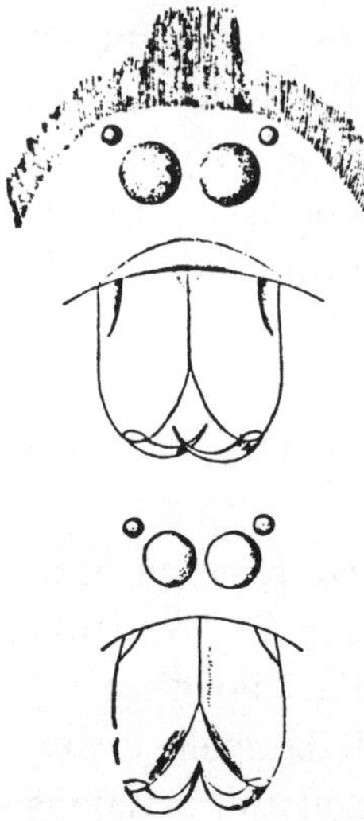

By far the most remarkable instance
of facial adornment is to be found in
Mopsus mormon[4] (Fig. 10.) The face of
the male is dark brown, covered with
steel-blue scales; around the eyes are
both brown and yellowish red hairs; the
front of the long falces is also dark
brown, covered with bright metallic
scales; and to add to his beauty there
is a high vertical ridge of variously tinted
hairs extending over his forehead.   The

Fig. 10.—Mopsus mormon   part above the middle eyes is brown, but
(from L. Koch). Upper figure
face and falces of male, show-   on either side it changes to pure white,
ing ridge of hairs; lower
figure, face and falces of   gradually becoming yellowish as it passes
female.
                                 back on to the sides.   In the female, while
the general coloration is the same, the most striking ornament—the band of hairs—is entirely absent.   In this species
we are reminded of the wonderful crests found in humming-
birds and fruit pigeons.   In *Dendryphantes elegans* the male alone
has two oblique converging ridges of short hairs extending from
the eyes of the second row to the anterior middle eyes; and in

1 Koch and Keyserling, *loc cit.*, p. 1173.
2 *Id ibid.*, p. 1181.
3 A. rufifrons, Simon: A. fusco manus, A. mystacalus, A. scops, Taczanowski.
4 *Arachniden Australiens*, p. 1319, described as *Ascyltus pencillatus.*

the black male form of *Astia vittata* there are three long tufts of black hairs on the eye-region, which are absent not only in the female but also in the other male form.

Leaving the *Attidæ*, there are several genera described by Menge, Cambridge, Simon, Thorell, Emerton and others that illustrate curious sexual modifications of the upper part of the face. For example, in the genus *Argyrodes*, (Fig. 11), of Simon, while the head of the female is often high and somewhat notched in front, in that of the male each of these divisions of the front part of the head gives rise to a horn, covered at the extremities with hairs. The drawing shows the parts in the two sexes. Emerton in *New England Therididæ*, when speaking of the genus *Ceratinella* says: "The heads of the males are usually higher than those of the females, and in some species are very large and raised into humps;" and of the genus *Cornicularia*

Fig. 11.—Argyrodes argyrodes (from Simon). Upper figure, cephalothorax of male; lower figure, cephalotharax of female.

he says that "the males have a hump or horn on the front of the head between the eyes, usually ornamented by flat, stiff hairs. In several species there are two horns, the lower one being small and partly concealed by the upper." There are many other genera in this family, with large numbers of species distributed over different parts of the world. Most of these species are small, but close inspection shows a great deal of sexual difference in the head parts and often also in the falces.

In male spiders the palpi are modified and serve as organs for the conveyance of the sperm cells to the epigynum of the female. Beside this direct use in the reproductive act they often play an important part as ornamental appendages. The female palpi, are, speaking generally, cylindrical, five-jointed outgrowths of the maxillæ, covered with hairs and varying somewhat in length and color; but throughout the order there is but little difference among the various species. Passing over the many and great differences of structure in the last or tarsal joint of the male palpus, which are primary sexual characters, we find, in this organ, many curious modifications of form and of ornamentation the only use of which is to please the female.[1]

In some species they are greatly elongated (*Laynus longimanus*); in others there are curious enlargements and apophyses on one or more of the joints (*Plexippus puerperus*); but the most frequent and by far the most striking form of decoration is a covering of long white or yellowish hairs which gives them a plume-like appearance (*Pensacola signata*).

In many spiders the sexes differ not only in the beautiful plumose palpi of the male, but also in the locomotive organs. The legs of the first pair are lengthened, in many males, and the several joints are enlarged and brilliantly colored, or furnished with long hairs or iridescent scale-like setæ. In no family of spiders does sexual selection seem to have been more effective than in the *Attidæ*. Many of the species, as we have seen, are furnished with remarkable falces, elaborate head ornaments and plume-like palpi, and we have now to give an account of a further modification which has apparently been gained for the sake of ornament, or possibly through the sham battles between the males.

In an ant-like species, *Synageles picata*, the female has legs of the ordinary form ; but in the male the tibia of the first leg is enlarged and flattened, and the anterior face of the enlargement is of a brilliant, steel-blue metallic color, as glossy as the breast of certain pigeons. In *Philæus metallescens*, from Australia, the legs of the first pair in the male are 11 mm. long, while those of the female are only 9 mm.; those of the male are of a very brilliant steel-blue color, and are ornamented with rings, spots and fringes of short, scale-like hairs, and of

---

1 Koch and Keyserling in *Arachniden Australiens* describe in the family *Attidæ* thirty-four males having well developed fringes or tufts of hair on the palpi, while there are only five females so ornamented, and several of these to only a moderate extent.

long, white hairs; those of the female are but slightly iridescent, and lack all further ornament. In *Dendryphantes elegans* the first leg in the male has on the lower half of the tibia a wide fringe of hair, which is very conspicuous as he waves his legs while courting; while the legs of the female are plain. These leg modifications have been frequently transmitted to the female, so that not uncommonly we find both sexes presenting striking forms, which are, however, proportionately less remarkable in the females than in the males. For example, in *Diolenius phrynoides,* where the first legs are wonderfully lengthened and modified, those of the female measure 12¼ mm., and those of the male 20½ mm., being nearly twice as long. The curious change in the legs of this species so impressed Walckenaer as to make him suggest that the spider must walk on the water, since in no other way could such legs be useful.

As it is a land species his explanation must be abandoned, and we are constrained to look upon these legs as secondary sexual organs, useless for locomotion, but of high importance while mating. To gain a clearer idea of this lengthening of the first legs one has only to imagine that in some group of human beings the arms of the men were doubled in length, while in the women they remained as before. There are numerous species in this genus, all characterized by their long legs (Fig. 12). It is not unusual for female as well as male beetles to possess well developed horns and knobs, so that there is nothing anomalous in the elongated legs of both sexes of the *Dio-*

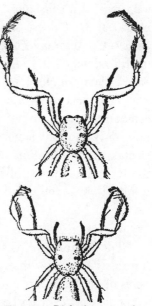

Fig. 12.—Diolenius venustus (from nature). Upper figure, male; lower figure, female.

lenii. In *Chirothecia,* a South American genus of *Attidæ,* we meet with other instances of modified legs, but here there is a marked difference between the sexes, the males having the legs longer, more robust, and more ornamented than the females. This is also true in the North American species *Icius palmarum*. *Saitis barbipes* has the third leg of the male fringed with hairs on either side, which gives it quite a plume-like appearance, while that of the female is entirely plain. In an undescribed species of *Habrocestum* from Arizona the first leg of the female is plain, while that of the male has on the tibia a fringe of long, silky, yellow hairs, mingled with which are other hairs, which are enlarged and flattened at the end. These spatulate hairs also appear on the first legs of *Habrocestum hirsutum*; in this species we have only the male.

Walckenaer remarks that his earlier division of *Attidæ* into *Sauteuses*, or short-legged, and *Voltigeuses*, or long-legged, is vicious, since in many species of *Sauteuses* the males have very long legs, and are, therefore, if we have not seen the female, placed in the *Voltigeuses*, while the females must be put into the *Sauteuses*.[1] We may go further than this, and say that all these modifications of the legs are sexual and of little or no importance in taxonomy.

The instances which we have given of secondary sexual differences might have been indefinitely multiplied, but we have only thought it necessary to give examples of each kind or class of modification; these will serve, we trust, to establish the fact that these differences are not less numerous among spiders than among birds and insects.

MATING HABITS.

For a number of years prior to 1888, we had been much impressed by the many important differences between the sexes of our jumping-spiders, and since we thought it most probable that they had come about through sexual selection, we had often tried to watch them during their courtship, but up to that time with very little success. We had occasional glimpses of their habits, but they were so incomplete as to con-

---

1 *Loc. cit.*, p. 482.

tribute but little to a thorough knowledge of the subject. Last year, we determined that, if possible, we would work out this subject so far as concerned the species in our locality. For this purpose we had made a number of mating-boxes. The larger ones were 15 inches long by 11½ wide and 3 deep; the smaller, 7¼ long by 5¾ wide and 2½ deep. The sides of each box were marked off into inches so that the distance of the spiders from each other could be easily noted. The floor was made of coarse cotton cloth, for the purpose of ventilation, while the top was of glass, so that the inmates of the cage could be kept fully in view at all times; this top could be opened and closed. As a usual thing we move into the country toward the last of June, but this year we went out on the 22d of May, in order to be in time for those species that mature early.

The courtship of spiders is a very tedious affair, going on hour after hour. We shall condense our descriptions as much as possible, but it must be noted that we often worked four or five hours a day for a week in getting a fair idea of the habits of a single species.

SAITIS PULEX.

On reaching the country we found that the males of *Saitis pulex* were mature and were waiting for the females, as is the way with both spiders and insects. In this species there is but little difference between the sexes. On May 24th, we found a mature female and placed her in one of the larger boxes, and the next day we put a male in with her. He saw her as she stood perfectly still, twelve inches away; the glance seemed to excite him and he at once moved toward her; when some four inches from her he stood still and then began the most remarkable performances that an armorous male could offer to an admiring female. She eyed him eagerly, changing her position from time to time so that he might be always in view. He, raising his whole body on one side by straightening out the legs, and lowering it on the other by folding the first two pairs of legs up and under, leaned so far over as to be in danger of losing his balance, which he only maintained by sidling rapidly toward the lowered side.   The palpus, too, on this side was turned back to correspond to the direction of the legs nearest it.   (Fig. 13.)   He moved in a semi-circle for about two inches and then instantly reversed the position of the legs and circled in the opposite direction, gradually approaching nearer and nearer to the female.   Now she dashes toward him, while he, raising his first pair of legs, extends them upward and forward as if to hold her off, but withal slowly retreats.   Again and

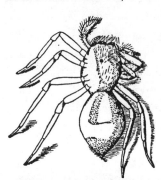

again he circles from side to side, she gazing toward him in a softer mood, evidently admiring the grace of his antics.   This is repeated until we have counted 111 circles made by the ardent little male.   Now he approaches nearer and nearer and when almost within reach whirls madly around and around her, she joining and whirling with him in a giddy maze.   Again he falls back and resumes his semi-circular motions,

Fig. 13. — Saitis pulex. Male dancing before female (from nature, by Mr. Ludwig Kumlien).

with his body tilted over; she, all excitement, lowers her head and raises her body so that it is almost vertical; both draw nearer; she moves slowly under him, he crawling over her head, and the mating is accomplished.

After they have paired once, the preliminary courtship is not so long.   When this same pair mated a second time, there were no whirling movements, nor did the female lift her body, as at first.   We watched this species a great deal during the three weeks that the mating lasted.   Once we saw a female approach a very glum-looking male, waving her palpi and making herself agreeable, but in vain.   He was pushed a little from without, so as to make him look toward her, when she turned about, holding her abdomen high and her head low.   Finally he grew excited enough to dance a little, and then they whirled round and round together in the usual manner ; but she appeared the more eager of the two.   This was true, however, of only this one female. All the others were more retiring, and were never more than half willing to be wooed.

The males, while very excitable in the presence of the female, do not seem to be especially quarrelsome. When excited, they pursue and leap upon each other, but do not exactly fight. During the display of the male before the female, she is no inattentive observer, but watches him intently, turning frequently to keep him in view as he dances from side to side, and, finally, if she approves of him, yields to his desires. Often, however, the male fails to make an impression upon her, even after dancing before her for a long time, since upon his too near approach, she runs away.

EPIBLEMUM SCENICUM.

On the afternoon of the 10th of June we found two males of this species fighting on a brick wall. They held up the first pair of legs and moved rapidly in front of each other, now advancing and now retreating, in a half circle, distant from each other about four and a half inches. There was little real earnestness in the affair, and it reminded one of the bluster of two boys, each threatening the other and daring the other to strike, but neither willing to be the aggressor. In a few minutes they both wandered away. During the next two days we found eighteen males and four females on this wall, which was about 20x12 feet in size. The performances of the male are not so complicated and interesting in this as in the preceding species, nor do the males seem to be so persistent.

We placed twelve spiders, of both sexes, in one of the boxes. Soon they were all moving about, the males making advances to the females, who seemed to endeavor to escape. After about two hours, we found that three pairs had come to an agreement and mated, the male, in each case, getting his female into a corner of the box and spinning a cover over and around her. Sometimes, while the male was working, the female would wander off several inches, but when the house was nearly completed, he would follow her, and half lead and half drive her home, when he would get her into the nest first, and then follow himself. Here the mating would be accomplished, after some slight preliminaries. The females seemed to have some difficulty in choosing from among the males, but, after a decision had been reached and a male accepted, there appeared to be complete agreement, and the male, thereafter, commenced to build his house. On the next morning we found all the pairs together in different tents.

The male, when prancing before the female, stood quite high on the three posterior pairs of legs; the first pair, the palpi, and the long falces, were stretched out stiffly in an oblique direction, the spider moving rapidly from side to side.

ICIUS SP.

This species, still undescribed, we discovered last summer, in large numbers, on rail fences at the edge of woods. We had worked over the place time and again, for many years past, but had overlooked them, not only on account of their admirable protective coloring, but from the fact that they only congregate, in this way, until they have mated, and afterward wander back into the woods and are then rarely met with. This is a habit with several *Attidæ*[1] and reminds one of the *"sícaleli"* or dancing-parties of certain birds, or of our own partridge dances. We were fortunate in discovering them just at the mating season. A dozen or more males and about half as many females were assembled together within the length of one of the rails. The males were rushing hither and thither, dancing opposite now one female and now another; often two males met each other, when a short passage of arms followed. They waved their first legs, sidled back and forth, and then rushed together and clinched, but quickly separated, neither being hurt, only to run off in search of other and fairer foes. We watched them for hours, and then, our patience being exhausted, filled all our bottles and carried them home. We placed them in the smaller boxes, since we have learned that propinquity is quite as effective in hastening the courtship of

---

1 In the genus *Dendryphantes, capitatus and elegans* are examples. Going to their favorite bushes we have caught 40–50 males in a few hours' sweeping, but after the season the same locality would not yield more than 2–3.

these little creatures as it is said to be among the higher races. For several days, in visiting this fence, we found goodly numbers on each rail, but after a little they grew scarcer and scarcer and at last we were unable to find one where a short time before they had been so common.

It was much more comfortable to study them after they had been put into the mating boxes and within the next few days we had seen many of them pair. The males were very quarrelsome and had frequent fights, but we never found that they were injured. Indeed, after having watched hundreds of seemingly terrible battles between the males of this and other species, the conclusion has been forced upon us that they are all sham affairs [1] gotten up for the purpose of displaying before the females, who commonly stand by, interested spectators. This is entirely contrary to what we had expected, and early in the season we, on several occasions, forcibly parted the combatants, fearing that they would kill each other. The falces in many of the males are, it is true, much lengthened, but as weapons of offense the shorter fangs of the female are much more deadly. In twelve species, in which we witnessed numberless fights, we could never discover that one of the valiant males was wounded in the slightest degree.

In this new species the position of the female while watching the male is unlike that of any of the others. She lies close to the ground with her first legs directed upward and forward, while her second legs are held on the ground and stretched forward in front of her face (Fig. 14). The male, when approaching her, does not throw his legs high over his head, as he does

Fig. 14.—Undescribed species. Position of female when approhched by male (from nature by L. K.)

before another male, but raises his body on his six hind legs;

---

1 Cuvier remarks that "the males sometimes engage in contests in which their manœuvres are very singular, but which do not terminate fatally." *Animal Kingdom,* trans. by Carpenter and Westwood, London, 1863. p. 464.

Hentz (*N. A. Spiders*, p. 133) saw two males of *Linyphia communis* fighting an

the first legs are held down toward the ground, diverging slightly near the head, and are bent inward at the middle so that the tips

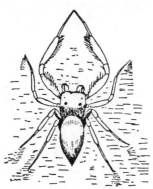

turn toward each other and meet. (Fig. 15.) At times he turns the apex of his abdomen down; at other times he keeps it straight, as he moves from side to side; the palpi are folded under. He sometimes varies his attitude by lying flat on his venter, keeping the tips of the legs touching as before.

Fig. 15.—Undescribed species. Position of male approaching female (from nature by L. K.).

HASARIUS HOYI.

The sexes are very different, so much so that we at first described them as two species, the male being the more conspicuous of the two. The males are ready in the early days of June, and the females a little later. In his dance the male has several movements; most commonly he goes rapidly from side to side with his first legs obliquely up; (Fig.

16); at other times he twists the abdomen to one side and bending low on the other, something as *pulex* did, goes first in one direction for about two inches, and then, reversing, circles to the opposite point. The females are very savage, especially with each other; and even the members of the sterner

Fig. 16.—Hasarius hoyi. Position of male approaching female (from nature by L. K.).

sex are not always free from danger when paying their preparatory addresses. Once we saw a female eagerly watching a prancing male and as he slowly approached her she raised her legs as if to strike him, but he, nothing daunted by her unkindly

obstinate battle; he did not watch it to the close, but believes that it was "without bloodshed."

Vinson, on the contrary, mentions a fatal combat between two males of *Epeira niger* that he had shut up in a bottle with a female. *Araneides de La Reunion, Maurices et Madagascar*, p. 190. We have a cousin of this species, *Argiope cophinaria*, and though we have seen the two or three little males that were courting a female manœuvre together the results were never serious.

reception of his attentions, advanced even nearer, when she seized him and seemed to hold him by the head for a minute— he struggling. At last he freed himself and ran away. This same male, after a time courted her successfully.

SYNAGELES PICATA.

These are small ant-like spiders. The most important sexual difference, is the greater thickness of the first legs of the male. These are flattened on the anterior surface and are of a brightly iridescent steel-blue color. Unlike most of the *Attid* males this species keeps all his feet on the ground during his courtship: raising himself on the tips of the posterior six he slightly inclines his head downward by bending his front legs, their convex surface being always turned forward. His abdomen is lifted vertically so that it is at a right angle to the plane of the cephalothorax. In this position he sways from side to side. After a moment he drops the abdomen, runs a few steps nearer the female, and then tips his body and begins to sway again. Now he runs in one direction, now in another, pausing every few moments to rock from side to side and to bend his brilliant legs so that she may look full at them. We were much impressed by the fact that the attitudes taken by the males served perfectly to show off their fine points to the female. We had never known the male of this species until the day that we caught this one and put him into the mating-box, and it was while studying his courtship that we noticed how he differed from the female in his iridescent first legs. He could not have chosen a better position than the one he took to make a display. We had six females in the box, and saw him mate with all of them ; and each, after a time, made a cocoon containing three large eggs.

MARPTUSA FAMILIARIS.

This is a rather plainly marked *Attus*, well protected by its coloring of gray and black, on the bark of trees and on fences, where it is most frequently found. There is little difference between the sexes. We placed the two together. She saw him as he entered at the opposite side of the box, some thirteen inches away. Eyeing him attentively, she slowly changed her position to keep him in sight, and kept her palpi moving rapidly, a characteristic action with the species. As he neared her, he stretched the first and second pairs of legs sideways, but, after a moment, backed away. (Fig. 17.) These manœuvres were repeated many times. Occasionally he would bend the tip of the abdomen down, lifting the body up on the last joints of the two hindermost legs. The female always paid the greatest attention to his movements, lying on the ground, with all the legs flattened out and the palpi slightly raised, the only movement visible being the vibration of the palpi. There is a certain slowness and dignity about the wooing of this species that is almost ludicrous.

Fig. 17.—Marptusa familiaris. Positions in courtship; left-hand figure female, right-hand figure male (from nature by L. K.).

PHIDIPPUS RUFUS.

The sexes are alike in general coloring, but the male has much the brighter tint. In the early days of July we found them mature and brought them together. The female standing still, the male, while some five inches away, stood high on the six back legs, turned the first pair forward and upward and crossed them at the tips; the palpi were held widely apart, parallel with the second legs. The effect of this position was to bring directly before her, as she stood watching him, the beautiful white hairs on the lower part of the palpi.[1] At the same time the abdomen dropped so that it touched the ground. In this way he advanced, with a swaying motion. The female ran away, but after a time he renewed his attentions. The

---

1. These white hairs, in contrast with the bright iridescent green falces, are very striking; the female, although her falces are green, is without the white hairs.

female *rufus* is a ferocious creature, having a great advantage in size, and so it happened that our assiduous male, in an unguarded moment, was pounced upon and eaten up.

PHIDIPPUS MORSITANS.

On this species we have but few notes. The single female that we caught during the summer was a savage monster. The two males that we provided for her had offered her only the merest civilities, when she leaped upon them and killed them. The sexes are quite alike in color and marking, but while the female has the fourth leg longest, the male has the first pair not only much the longer, but thickly adorned with white hairs, some of which are long and others short and scale-like. It was while one of the males was waving these handsome legs over his head that he was seized by his mate and devoured. The tibia of the palpus is also covered with white hairs, which make a strong contrast with the general black color, and this is held out in such a way as to make a display as he approaches the female.

DENDRYPHANTES CAPITATUS.

The sexes are entirely different. In the male the bronze brown face is made very conspicuous by some snowy white bands, as is shown in the drawing. These are wanting in the female, her face being rufus with some few scattering white hairs. The males of *capitatus* are very quarrelsome, sparring whenever they meet, chasing each other about, and sometimes clinching. It is a very abundant spider with us, so that we often put eight or ten males into a box to see them fight. It seemed cruel sport at first, but it was soon apparent that they were very prudent little fellows, and were fully conscious that "he who fights and runs away will live to fight another day." In fact, after two weeks of hard fighting we were unable to discover one wounded warrior. When the males are approaching each other, they hold the first legs up in a vertical direction. Sometimes they drop the body on to one side as they jump about each other. These movements are very quick, and they

are always ready for a passage at arms. When courting the
females they have another movement. They approach her
rapidly until within two to five inches, when they stop and extend the first legs directly forward, close to the ground, the legs

being slightly curved with the
tips turned up. (Fig. 18.)
Whether it be intentional or
not, this position serves admirably to expose the whole of the
bronze and white face to the
attentive female, who watches

Fig. 18.—Dendryphantes capitatus.
Position of male approaching female
(from nature, by L. K.).

him closely from a little distance. (Fig. 19.) The males also
give their palpi a circular movement, much as a person does
when washing his hands. As he grows more excited, he lies

down on one side with his legs still extended.
These antics are repeated for a very long time,
often for hours, when at last the female, either
won by his beauty or worn out by his persis-
tence, accepts his addresses.

Fig. 19.—Dendry-
phantes capitatus.
Face and palpi of
male (from nature,
by L. K.).

DENDRYPHANTES ELEGANS.

The male of the species, like many other
animals, has received a number of names.
Hentz called the female *elegans*, and the male
superciliosus. C. Koch called him *cristata*, and we ourselves,
ibialis, on account of the fringe of hairs on the tibia of his
first leg. Both sexes are beautiful. The male is covered
with iridescent scales, his general color being green; in the
female the coloring is dark, but iridescent, and in certain lights
has lovely rosy tints. In the sunlight both shine with the
metallic splendor of humming-birds. The male alone has a
superciliary fringe of hairs on either side of his head, his first
legs being also longer and more adorned than those of his
mate. The female is much larger, and her loveliness is
accompanied by an extreme irritability of temper which the
male seems to regard as a constant menace to his safety, but
his eagerness being great, and his manners devoted and tender,

he gradually overcomes her opposition. Her change of mood is only brought about after much patient courting on his part. While from three to five inches distant from her he begins to wave his plumy first legs in a way that reminds one of a wind-mill. She eyes him fiercely and he keeps at a proper distance for a long time. If he comes close she dashes at him and he quickly retreats. Sometimes he becomes bolder and when within an inch, pauses, with the first legs outstretched before him, not raised as is common in other species; the palpi also are held stiffly out in front with the points together. Again

Fig. 20.—Zygoballus bettini. Position of male approaching female (from nature, by L. K.).

she drives him off, and so the play continues. Now the male grows excited as he approaches her, and while still several inches away whirls completely around and around; pausing, he runs closer and begins to make his abdomen quiver as he stands on tip-toe in front of her. Prancing from side to side, he grows bolder and bolder, while she seems less fierce, and yielding to the excitement lifts up her magnificently iridescent abdomen, holding it at one time vertically and at another sideways to him. She no longer rushes at him, but retreats a little as he approaches. At last he comes close to her, lying flat, with his first legs stretched out and quivering. With the tips of his front legs he gently pats her; this seems to arouse the old demon of resistance, and she drives him back. Again and again he pats her with a caressing movement, gradually creeping nearer and nearer, which she now permits without resistance until he crawls over her head to her abdomen, far enough to reach the epigynum with his palpus.

ZYGOBALLUS BETTINI.

The sexual differences in this species are well marked. The male has much more silvery white on the face; the first legs and falces are much longer, and on the side of the abdomen the general color is darker and the lateral bars of a much more glistening white. All the colors are far more brilliant in the male than in the female. (Plate II.) In courting the male lies flat near the female, wriggling his abdomen and frequently turning from side to side. His first legs are held up over his head, slightly diverging, and are often twisted and turned about. (Fig. 20, see p. 47.)

Two males that were displaying before one female, rushed savagely upon each other and fought for twenty-two minutes, during one round remaining clinched for six minutes. When fighting, the abdomen is held nearly at a right angle with the

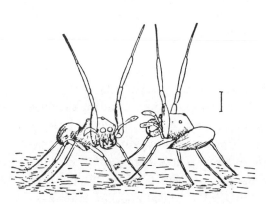

Fig. 21.—Zygoballus bettini. Position of males when fighting (from nature, by L. K.).

c e p h a l o t horax. (Fig. 21.) T h e c o m b a tants appeared tired at the close of the battle, but after a short rest were perfectly well and fought a number of times s u b s e q u e n t l y. There a r e t w o forms of male in this species, o n e b e i n g twice as large as the other.

HABROCESTUM SPLENDENS.

The colored plate gives a good idea of the sexual difference in this species. The male, a magnificent fellow, when we first caught him, displayed for a long time before the female. He began by advancing a few inches toward her and then backing off again, this being repeated many times. After a while he settled down under a little web in the corner. The female, troubled by this indifferent treatment, advanced toward him; he came out and she fell back. This play was kept up for some time, but at length the male began his courting in earnest. When within a few inches of her he began a rapid dance from side to side, raising the whole body high on the tips of the legs, the first pair being directed forward and the palpi clasped together, with the abdomen turned to one side and lifted up. After a short dance he stood motionless, striking an attitude, as shown in the figure, remaining quiet for half a

minute. (Fig. 22.) Then he turned his back on her, moving irregularly a b o u t with his legs forward and his palpi vibrat- ing. Again he dances sideways before her, strutting and show- ing off like a pea-

Fig. 22.—Habrocestum splendens. Position of male approaching female (from nature, by L. K.).

cock, or whirling around and around. When he turned his back we often thought that the female seemed disappointed, since she would then commonly move nearer to him and ap- pear much excited herself. We at first supposed that this turning around was accidental, but it happened so regularly at a certain stage of the courtship that we concluded it was an important part of his display, serving to better show off his brilliant abdomen. Our artist, Mr. Kumlien, while watching the antics in order to draw the spider, called our attention to this habit, not knowing that we had observed it. The fact that among spiders the males take such attitudes as display their best points recalls this passage in one of Darwin's letters: "I am very glad to hear of your cases of the two sets of *Hesper- iadæ*, which display their wings differently, according to which surface is colored. I cannot believe that such display is accidental or purposeless."

ICIUS MITRATUS.

The male is quite different from the female, especially in his slender, tapering body and in his long first legs. While in *splendens* the female was remarkable for the attention that she gave the male, seeming at times to coquet with him, in this species she was remarkable for her indifference. She takes less interest in the display of the male than any spider that has come under our observation. Another peculiarity of *mitratus* is the large amount of time and strength given up by the males to fighting. They do not seem so fierce as other species, but they cannot endure each other's proximity. A male will leave a female in the midst of his caresses to drive off another male that comes too near. We once saw a male jump on to the back of another that was pairing with a female. The latter turned on the intruder and drove him to a distance of five or six inches, and then returned and renewed his addresses. In courting and in fighting the position of the male is the same. The body is somewhat raised; the first legs are held at a right angle to the cephalothorax, and the abdomen is twisted to one side, and, as he dances before the female, is changed now to the

Fig. 23.—Icius mitratus. Male dancing before female (from nature, by L. K.).

right, now to the left. (Fig. 23.) In mating, the male does not usually crawl over her, but jumps upon her from a distance of one or two inches, in this respect agreeing with the following species, *militaris*. Whether or no this habit in the two species is due to the savageness of the female—the females of both species sometimes attack the males—we are unable to say.

PHILÆUS MILITARIS.

This is the only species in which we saw males take possession of young females and keep guard over them until they became mature. There is a good deal of difference in the size of the males, some being larger and some smaller than the females. We commenced our experiments by putting half a dozen mature males into a box. They at once began to chase each other about and to threaten each other with upraised legs.

The addition to the company of a number of nearly mature females considerably increased this tendency to quarrelsomeness, the males thereafter spending all the time that they spared from courtship in fighting. In approaching the female they seemed very eager, and fairly quivered with excitement. The first two legs were raised over the head and curved toward each other, so that the tips nearly met and the palpi were moved up and down. (Fig. 24.) One of them was much larger than any of the others, and from the first showed plainly that he had a strong sense of his own superiority. He drove all the others about at his pleasure, constantly interrupting their attempts at courtship. This big fellow seemed to be especially attractive to the females, of which there were always two or three standing in an admiring circle around him. When he approached them, however, they slipped away. After a time, he singled

Fig. 24.—Philæus militaris. Position of male approaching female (from nature, by L. K.).

out the largest of the females, half coaxed and half drove her into a corner, and there kept her secluded, chasing away every spider that approached, irrespective of sex. He once interrupted a courtship which was going on six inches away, driving the male to a distance and then pursuing the female for a long time. This seclusion of the female was kept up until evening of the next day—a period of twenty-four hours—but on the following morning the pair had separated and he was hidden among some leaves. None of the females were yet mature. We now put into the box another male, which was nearly as large as the first. This second one adopted the same bullying manner to the smaller members of his sex that the first had done. After driving them about for a time, he secured a gnat, and was peacefully devouring it, when number one emerged from the leaves, caught sight of the new-comer, and at once approached, bristling with pride and ire, his first legs raised high, as if to strike, his palpi vibrating with excitement, and his abdomen dragging, first on one side and then on the other. Number two was evidently of good courage, for he held his ground, and, not relinquishing the gnat, raised his legs and clinched with his antagonist. The battle raged for five minutes; finally number one pulled the gnat from the other, and then chased him away.

For several days following life in the mating box was robbed of its monotony by perpetual battles among the males. The females, in eluding them, jumped and hung from a thread. At one time a small one guarded a female for some hours in a corner; she once slipped away and ran a few inches, but when another male began to pay court to her, she ran back and crept under the body of her protector. After two or three days each of the two large males took possession of a female, spun a web over her, and spinning a second sheet above as a cover for himself, remained quiet in the little nest thus formed for a week. During this time every spider that approached was driven away. They went out occasionally for food, but were not seen to carry any to their mates.

At the end of a week number one was observed to be pairing with his female, which had moulted and was now mature. The two were separated, when the male went hunting about and finally sprang upon an unmated female, three inches away from the other nest, whereupon number two ran out, attacked him violently, drove him away, and then returned to his nest. Number one, while wandering about, caught sight of his own mate and sprung upon her without any of the preliminary attention before noticed. From this time forward, the big, wandering male, his occupation gone, became a very thorn in the flesh to the other, whose female had not yet matured. Not only did he continually approach the nest, thus arousing a jealous fury in its owner, which was not called out by any of the other spiders, but whenever the rightful owner was away from home, chasing intruders or procuring food, this disturber of domestic peace made his way into the nest. The first time this happened, the owner, returning, ejected him without ceremony; the second time they had a prolonged struggle, clinching, and falling, thus locked together, a distance of about twelve inches—the height of the box (Fig. 25); the third time number one was not discovered until he had cut the still immature female out of the web which enclosed her. She ran away, and after that the two males wandered about, fi g h t i n g whenever they met. The defrauded male, as well as the other one, now

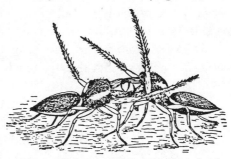

Fig. 25.—Philæus militaris. Position of males when fighting (from nature, by L. K.).

courted every female in the box, although so long as they had their own mates they had paid no attention to other females, except to drive them away.

The following extract from our notes shows that the guarding of young females is a habit of the males in this species, and was not the result of artificial conditions:

"Aug. 17. Found a mature male *militaris* standing guard, in a tent, over an immature female (one moult from maturity). They seemed very friendly when taken out and put into a bottle. The webs and positions were as we had seen them in our mating-boxes.

ASTIA VITTATA.

There is a good deal of interest connected with the study of this species, for the reason that there are two well marked male forms; moreover, their love antics are unusually curious. A description of the two males is unnecessary, since they are well represented on Plate II. The two forms grade into each other, excepting that the three hair tufts are only found in the fully developed *niger* form. The *vittata* form, which is quite like the female, when he approaches her, raises his first legs either so that they point forward or upward, keeping his palpi stiffly outstretched, while the tip of his abdomen is bent to the ground.   This position he commonly takes when three or four inches away.   While he retains this attitude he keeps curving and waving his legs in a very curious manner.   Frequently he raises only one of the legs of the first pair, running all the time from side to side.   As he draws nearer to the female he lowers his body to the ground, and, dropping his legs also, places the two anterior pairs so that the tips touch in front, the proximal

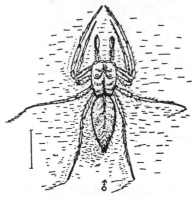

joints being turned almost at a right angle to the body. (Fig. 26.) Now he glides in a semi-circle before the female, sometimes advancing, sometimes receding, until at last she accepts his addresses. The *niger* form, evidently a later development, is much the more lively of the two, and whenever the two varieties were seen to compete for a female, the black one was suc-

Fig. 26.—Astia vittata. Position of male approaching female (from nature, by L. K.).

cessful.   He is bolder in his manners, and we have never seen him assume the prone position, as the red form did, when close to the female.   He always held one or both of the first legs high in the air, waving them wildly to and fro, or, when the female became excited, he stood perfectly motionless before her, sometimes for a whole minute, seeming to fascinate her by the power of his glance.   (Fig. 27, see p. 55.)   A female that was full of eggs was looked at critically from a distance of four or five inches by several ardent males, but received no further attention.   Although the males were continually waving their first legs at each other, their quarrels were harmless.   It was quite otherwise with the females, since they not only kept the other sex in awe of them, but not infrequently, in their battles, killed each other.   We thought it rather remarkable that the *niger* variety should not have the same antics as the *vittata;* but since closely related species often differ greatly in color, form

and habits, we should not be surprised at finding like differences between distinct varieties.

The very meager knowledge that we have of the mating
habits of the *Epeiridæ* has been gained at the expense of numerous hours of watching.   The courtship of *Attidæ* is often very
tedious, but it does not compare in this respect with that of the
Epeiridæ, perhaps because the members of the former family
are constantly in motion and thus hold out a hope to the observer that he is going to see something worth seeing, while
the orb-weavers make nothing of remaining motionless for six
or eight hours at a stretch; the observer, in the meantime, being afraid to let his attention flag
for a moment lest he lose some
small but significant movement
on their part.   We have watched
the males and females of several
species, but our only notes per-
tain to two species of *Argiope*,
cophinaria and *fasciata*, in which
the males are much smaller than
the females.   In the mating season each female has three or four
of these little males hanging
about the outskirts of her web.

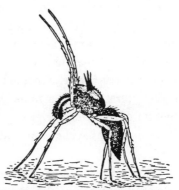

Fig. 27.— Astia vittata, var. niger.
Position of male approaching female
(from nature, by L. K.).

In *cophinaria*, when two of these males meet, they throw
up their first legs and back away from each other, without
striking or clinching as in the *Attidæ*, and then one of them
drops at the end of a line.   When advancing toward the
female, the male seems to pause and pull at the strands of web,
as though to notify her of his approach.   When he comes
toward her from in front she imparts a slight motion to the
web with her legs, which seems to serve as a warning, as he
either moves away or drops out of the web.   When he comes
from behind she pays no attention to him until he begins to
creep on to her body, when she slowly raises one of her long
legs and unceremoniously brushes him off.

We passed one afternoon in watching a female of *fasciata* in her web, around the edge of which were perched four little males. The proceedings were briefly as follows:

One of the males ran lightly over the web toward the female, approaching her from behind. Before he reached her she, seemingly conscious of his approach, gave the web a violent shake, whereupon he retreated. He made three or four trials of this kind and then seemed to give up hope. His place was taken by one of the other males, which acted in exactly the same way, with no greater success, and then gave place to a third. The female was always approached from above and behind, she hanging head downward, and she usually gave her warning shake before the male came very near, although once or twice he came close enough to touch her. The males showed no ill feeling toward each other. The most interesting part of it was that she seemed to recognize from the character of the vibration that a male was approaching, not taking the motion of the web to signify that an insect was entangled in it, as in this case she would at once have turned to secure her prey.

From these slight observations we were inclined to believe that the courtship in the *Epeiridæ* was carried on, to some extent at least, by a vibration of web lines. Dr. McCook subsequently confirmed that opinion, and we quote from his work on *American Spiders and their Spinning Industry* the following extract, which bears directly on the subject :

'The first stages of courtship have already been indicated. Having found the snare of his partner, the male stations himself upon the outer border and awaits results. It is not difficult for him to communicate his presence. Indeed, he must take his place deftly and keep it very quietly upon the snare, or he will quickly bring down upon him the voracious lady of the house. A touch of his claw upon a radius would telegraph to the female the fact of his presence; and I believe, from what I have seen of the operations of the male in this preliminary stage of courtship, as well as from the recorded observations of others, that he does thus intimate his presence, and that the first stages of the engagement are consummated by these telegraphic communications back and forth between male and female over the delicate filaments of the silken snare.

"If matters be favorable, the male draws nearer, usually by short approaches, renewing the signals at the halting places. Sometimes this preliminary stay is very brief; sometimes it is greatly prolonged. I have observed it to be continued during several days, in which the male would patiently wait, sometimes, but not always, changing his position until his advances were favorably received, or were so decidedly repulsed that he was compelled to retire. With Labyrinth spiders I have generally seen the male stationed upon the maze, or that part of the snare which consists of crossed lines. Here he would make for himself, as he hung back downward, a little dome of spinning work, which spread above him like a miniature umbrella. The male of *Argiope cophinaria* feels the web with his feet for some time before the final approach. The male of *Linyphia marginata*, as he cautiously approaches, pulls upon the threads connecting his own with his lady's bower. The period of approach or courtship is generally terminated by a sudden rush which brings the partners into union."

The same idea was advocated many years ago by Termeyer, as follows: [1]

"The *diadema* spider was that which I examined. * * * He never appears in the center of the beautiful webs, and even when I saw him he was, as to abdomen and palpi, so different from the female, which in other respects he resembled, that I should not have supposed him of the same species. He never spins webs except in the time of his amours. * * * He approaches ltttle by little, with much caution, doubtful of the reception with which he is to meet in the web of the female, who occupies the center, intent only on her prey. He commences

1 Researches and Experiments upon the Silk from Spiders and upon their Reproduction. Raymond Maria de Termeyer. Revised by Burt G. Wilder, M. D. Proceedings of Essex Institute, Vol. V, pp. 71–73.

by touching with one leg a thread of her web; the female approaches him; he flies, allowing himself to hang. Then he rises, winding up the thread, when he is assured, by I know not what movement, that he will not be ill-received; then he approaches her and with one of the palpi touches her stomach quickly many times. Then he returns, repeats the same act and departs, if he succeeds in leaving. I say if he succeeds, because I wish to relate what came under my observation in 1798. * * * I put a male and female of the *diadema* spider together in a box like a drum, closed with a veil at both ends. The male began by making various movements, as if to draw the attention of the female, who pretended not to perceive him, but only from time to time touched some thread of the web. He boldly approached, directing one of his palpi to her abdomen, and she extending this toward the palp. * * * But I saw also with surprise and indignation that, the work hardly finished, the male not being able to fly on account of the confinement, the female enveloped him in her thread, and having thus deprived him of every means of defense, devoured him. Perhaps overpowering hunger compelled her to it, but the act was very ferocious."

SUMMARY AND CONCLUSION.

In this paper we have considered the two theories by which Mr. Wallace explains sexual color differences in animals—primarily the greater vitality of the males, especially during the breeding season, and secondarily the greater need of protective coloring on the part of the females; and we have found that however satisfactory they may be where birds and butterflies are concerned, they fail in each important particular when applied to spiders.

In our study of moulting habits we have seen that among the *Attidæ*, where the sexual differences are strongest, males are commonly more brilliant than the females; that the young males nearly always resemble the adult females; that the males, when they differ from the females, depart from the general coloring of the group; and that when the females depart from the coloring of the group they approach, in the same degree, the coloring of the males. Mr. Wallace's theory would only partially explain these facts since although the increased vitality at the breeding season might produce variations which would tend to be inherited at that age, the assumption with which he starts out—that the male animal is constitutionally more active than the female—is not true in regard to spiders.

While studying the secondary sexual characteristics of spiders we came upon several large groups of facts which seem entirely inconsistent with Mr. Wallace's view. First, we found no evidence that the male spiders possess greater vital activity; on the contrary, it is the female that is the more active and pugnacious of the two. Second, we found no relation, in either sex, between development of color and activity; the *Lycosidæ*, which are among the most active of all spiders, having the least color development, while the sedentary orb-weavers show the most brilliant hues. Third, we found that in the numerous cases where the male differed from the female by brighter colors and ornamental appendages, these adornments were not only so placed as to be in full view of the female during courtship, but that the attitudes and antics of the male spider at that time were actually such as to display them to the fullest extent possible. Moreover, we noticed that the males were much more quarrelsome in the presence of the females, and that they, to a great extent, lost their tendency to fight when the mating season was over.

With these facts in mind let us examine Mr. Wallace's two strongest objections to the theory of sexual selection.

First: " There is a total absence of any evidence that the females admire or even notice the display of the males. Among butterflies there is literally not one particle of evidence that the female is influenced by color or even that she has any power of choice, while there is much direct evidence to the contrary."[1] In butterflies and in birds, with their rapid flight, it is difficult to determine how much one sex is watched by the other; but in the *Attidæ* we have conclusive evidence that the

---

1 *Tropical Nature*, pp. 199–200.

females pay close attention to the love dances of the males, and also that they have not only the power, but the will, to exercise a choice among the suitors for their favor.

Second: The fact that every male bird finds a mate "would almost or quite neutralize any effect of sexual selection of color or ornament; since the less highly colored birds would be at no disadvantage as regards leaving healthy offspring." In spiders, as the females gradually become adult, they have a choice from among a number of males, as these mature several days earlier. The males will pair as often as they have the opportunity and as the mating season lasts for two or three weeks the more brilliant males may easily be selected again and again.

The fact that in the *Attidæ* the males vie with each other in making an elaborate display, not only of their grace and agility but also of their beauty, before the females, and that the females, after attentively watching the dances and tournaments which have been executed for their gratification, select for their mates the males that they find most pleasing, points strongly to the conclusion that the great differences in color and in ornament between the males and females of these spiders are the result of sexual selection.

NOTE.—Since finishing the above we have seen, in the February number of the Popular Science Monthly, T. H. Morgan's article on *The Dance of the Lady Crab*. The observation therein noted is full of interest, showing, as it does, that sexual display in the invertebrates is not confined to spiders.

22

Reprinted from pages 273 and 282–298 of *Darwinism: An Exposition of the Theory of Natural Selection with some of Its Applications,* Macmillan and Co., London and New York, 1889, 494p.

COLOURS AND ORNAMENTS CHARACTERISTIC OF SEX

A. R. Wallace

[*Editor's Note:* In the original, material precedes these excerpts.]

Probable Causes of these Colours.

In the production of these varied results there have probably been several causes at work. There seems to be a constant tendency in the male of most animals—but especially of birds and insects—to develop more and more intensity of colour, often culminating in brilliant metallic blues or greens or the most splendid iridescent hues; while, at the same time, natural selection is constantly at work, preventing the female from acquiring these same tints, or modifying her colours in various directions to secure protection by assimilating her to her surroundings, or by producing mimicry of some protected form. At the same time, the need for recognition must be satisfied; and this seems to have led to diversities of colour in allied species, sometimes the female, sometimes the male undergoing the greatest change according as one or other could be modified with the greatest ease, and so as to interfere least with the welfare of the race. Hence it is that sometimes the males of allied species vary most, as in the different species of Epicalia; sometimes the females, as in the magnificent green species of Ornithoptera and the "Æneas" group of Papilio.

The importance of the two principles—the need of protection and recognition—in modifying the comparative coloration of the sexes among butterflies, is beautifully illustrated in the case of the groups which are protected by their distastefulness, and whose females do not, therefore, need the protection afforded by sober colours.

[*Editor's Note:* Material has been omitted at this point.]

Sexual Selection by the Struggles of Males.

Among the higher animals it is a very general fact that the males fight together for the possession of the females. This leads, in polygamous animals especially, to the stronger or better armed males becoming the parents of the next generation, which inherits the peculiarities of the parents; and thus vigour and offensive weapons are continually increased in the males, resulting in the strength and horns of the bull, the tusks of the boar, the antlers of the stag, and the spurs and fighting instinct of the gamecock. But almost all male animals fight together, though not specially armed; even hares, moles, squirrels, and beavers fight to the death, and are often found to be scarred and wounded. The same rule applies to almost all male birds; and these battles have been observed in such different groups as humming-birds, finches, goatsuckers, woodpeckers, ducks, and waders. Among reptiles, battles of the males are known to occur in the cases of crocodiles, lizards, and tortoises; among fishes, in those of salmon and sticklebats. Even among insects the same law prevails; and male spiders, beetles of many groups, crickets, and butterflies often fight together.

From this very general phenomenon there necessarily results a form of natural selection which increases the vigour and fighting power of the male animal, since, in every case, the weaker are either killed, wounded, or driven away. This selection would be more powerful if males were always in excess of females, but after much research Mr. Darwin could not obtain any satisfactory evidence that this was the case. The same effect, however, is produced in some cases by constitution or habits; thus male insects usually emerge first from the pupa, and among migrating birds the males arrive first both in this country and in North America. The struggle is thus intensified, and the most vigorous males are the first to have offspring. This in all probability is a great advantage, as the early breeders have the start in securing food, and the young are strong enough to protect themselves while the later broods are being produced.

It is to this form of male rivalry that Mr. Darwin first applied the term "sexual selection." It is evidently a real power in nature; and to it we must impute the development of the exceptional strength, size, and activity of the male, together with the possession of special offensive and defensive weapons, and of all other characters which arise from the development of these or are correlated with them. But he has extended the principle into a totally different field of action, which has none of that character of constancy and of inevitable result that attaches to natural selection, including male rivalry; for by far the larger portion of the phenomena, which he endeavours to explain by the direct action of sexual selection, can only be so explained on the hypothesis that the immediate agency is female choice or preference. It is to this that he imputes the origin of all secondary sexual characters other than weapons of offence and defence, of all the ornamental crests and accessory plumes of birds, the stridulating sounds of insects, the crests and beards of monkeys and other mammals, and the brilliant colours and patterns of male birds and butterflies. He even goes further, and imputes to it a large portion of the brilliant colour that occurs in both sexes, on the principle that variations occurring in one sex are sometimes transmitted to the same sex only, sometimes to both, owing to peculiarities in the laws of inheritance. In this extension of sexual selection to include the action of female choice or preference, and in the attempt to give to that choice such wide-reaching effects, I am unable to follow him more than a very little way; and I will now state some of the reasons why I think his views are unsound.

Sexual Characters due to Natural Selection.

Besides the acquisition of weapons by the male for the purpose of fighting with other males, there are some other sexual characters which may have been produced by natural selection. Such are the various sounds and odours which are peculiar to the male, and which serve as a call to the female or as an indication of his presence. These are evidently a valuable addition to the means of recognition of the two sexes, and are a further indication that the pairing season has arrived; and the production, intensification, and differentiation of these sounds and odours are clearly within the power of natural selection. The same remark will apply to the peculiar calls of birds, and even to the singing of the males. These may well have originated merely as a means of recognition between the two sexes of a species, and as an invitation from the male to the female bird. When the individuals of a species are widely scattered, such a call must be of great importance in enabling pairing to take place as early as possible, and thus the clearness, loudness, and individuality of the song becomes a useful character, and therefore the subject of natural selection. Such is especially the case with the cuckoo, and with all solitary birds, and it may have been equally important at some period of the development of all birds. The act of singing is evidently a pleasurable one; and it probably serves as an outlet for superabundant nervous energy and excitement, just as dancing, singing, and field sports do with us. It is suggestive of this view that the exercise of the vocal power seems to be complementary to the development of accessory plumes and ornaments, all our finest singing birds being plainly coloured, and with no crests, neck or tail plumes to display; while the gorgeously ornamented birds of the tropics have no song, and those which expend much energy in display of plumage, as the turkey, peacocks, birds of paradise, and humming-birds, have comparatively an insignificant development of voice. Some birds have, in the wings or tail, peculiarly developed feathers which produce special sounds. In some of the little manakins of Brazil, two or three of the wing-feathers are curiously shaped and stiffened in the male, so that the bird is able to produce with them a peculiar snapping or cracking sound; and the tail-feathers of several species of snipe are so narrowed as to produce distinct drumming, whistling, or switching sounds when the birds descend rapidly from a great height. All these are probably recognition and call notes, useful to each species in relation to the most important function of their lives, and thus capable of being developed by the agency of natural selection.

Decorative Plumage of Birds and its Display.

Mr. Darwin has devoted four chapters of his *Descent of Man* to the colours of birds, their decorative plumage, and its display at the pairing season; and it is on this latter circumstance that he founds his theory, that both the plumage and the colours have been developed by the prefer ence of the females, the more ornamented males becoming the parents of each successive generation. Any one who reads these most interesting chapters will admit, that the fact of the display is demonstrated; and it may also be admitted, as highly probable, that the female is pleased or excited by the display. But it by no means follows that slight differences in the shape, pattern, or colours of the ornamental plumes are what lead a female to give the preference to one male over another; still less that all the females of a species, or the great majority of them, over a wide area of country, and for many successive generations, prefer exactly the same modification of the colour or ornament.

The evidence on this matter is very scanty, and in most cases not at all to the point. Some peahens preferred an old pied peacock; albino birds in a state of nature have never been seen paired with other birds; a Canada goose paired with a Bernicle gander; a male widgeon preferred a pintail duck to its own species; a hen canary preferred a male greenfinch to either linnet, goldfinch, siskin, or chaffinch. These cases are evidently exceptional, and are not such as generally occur in nature; and they only prove that the female does exert some choice between very different males, and some observa tions on birds in a state of nature prove the same thing; but there is no evidence that slight variations in the colour or plumes, in the way of increased intensity or complexity, are what determines the choice. On the other hand, Mr. Darwin gives much evidence that it is *not* so determined. He tells us that Messrs. Hewitt, Tegetmeier, and Brent, three of the highest authorities and best observers, "do not believe that the females prefer certain males on account of the beauty of their plumage." Mr. Hewitt was convinced " that the female almost invariably prefers the most vigorous, defiant, and mettlesome male;" and Mr. Tegetmeier, "that a gamecock, though disfigured by being dubbed, and with his hackles trimmed, would be accepted as readily as a male retaining all his natural ornaments."[1] Evidence is adduced that a female pigeon will sometimes take an antipathy to a particular male without any assignable cause; or, in other cases, will take a strong fancy to some one bird, and will desert her own mate for him; but it is not stated that superiority or inferiority of plumage has anything to do with these fancies. Two instances are indeed given, of male birds being rejected, which had lost their ornamental plumage; but in both cases (a widow-finch and a silver pheasant) the long tail-plumes are the indication of sexual maturity. Such cases do not support the idea that males with the tail-feathers a trifle longer, or the colours a trifle brighter, are generally preferred, and that those which are only a little inferior are as generally rejected,—and this is what is absolutely needed to establish the theory of the development of these plumes by means of the choice of the female.

It will be seen, that female birds have unaccountable likes and dislikes in the matter of their partners, just as we have ourselves, and this may afford us an illustration. A young man, when courting, brushes or curls his hair, and has his moustache, beard, or whiskers in perfect order, and no doubt his sweetheart admires them; but this does not prove that she marries him on account of these ornaments, still less that hair, beard, whiskers, and moustache were developed by the continued preferences of the female sex. So, a girl likes to see her lover well and fashionably dressed, and he always dresses as well as he can when he visits her; but we cannot conclude from this that the whole series of male costumes, from the brilliantly coloured, puffed, and slashed doublet and hose of the Elizabethan period, through the gorgeous coats, long waistcoats, and pigtails of the early Georgian era, down to the funereal dress-suit of the present day, are the direct result of female preference. In like manner, female birds may be

[1] *Descent of Man*, pp. 417, 418, 420.

charmed or excited by the fine display of plumage by the males; but there is no proof whatever that slight differences in that display have any effect in determining their choice of a partner.

Display of Decorative Plumage.

The extraordinary manner in which most birds display their plumage at the time of courtship, apparently with the full knowledge that it is beautiful, constitutes one of Mr. Darwin's strongest arguments. It is, no doubt, a very curious and interesting phenomenon, and indicates a connection between the exertion of particular muscles and the development of colour and ornament; but, for the reasons just given, it does not prove that the ornament has been developed by female choice. During excitement, and when the organism develops superabundant energy, many animals find it pleasurable to exercise their various muscles, often in fantastic ways, as seen in the gambols of kittens, lambs, and other young animals. But at the time of pairing, male birds are in a state of the most perfect development, and possess an enormous store of vitality; and under the excitement of the sexual passion they perform strange antics or rapid flights, as much probably from an internal impulse to motion and exertion as with any desire to please their mates. Such are the rapid descent of the snipe, the soaring and singing of the lark, and the dances of the cock-of-the-rock and of many other birds.

It is very suggestive that similar strange movements are performed by many birds which have no ornamental plumage to display. Goatsuckers, geese, carrion vultures, and many other birds of plain plumage have been observed to dance, spread their wings or tails, and perform strange love-antics. The courtship of the great albatross, a most unwieldy and dull coloured bird, has been thus described by Professor Moseley: "The male, standing by the female on the nest, raises his wings, spreads his tail and elevates it, throws up his head with the bill in the air, or stretches it straight out, or forwards, as far as he can, and then utters a curious cry."[1] Mr. Jenner Weir informs me that "the male blackbird is full of action, spreads out his glossy wing and tail, turns his rich golden

[1] *Notes of a Naturalist on the Challenger.*

beak towards the female, and chuckles with delight," while he
has never seen the more plain coloured thrush demonstrative
to the female.   The linnet distends his rosy breast, and
slightly expands his brown wings and tail; while the various
gay coloured Australian finches adopt such attitudes and
postures as, in every case, to show off their variously coloured
plumage to the best advantage.[1]

A Theory of Animal Coloration.

Having rejected Mr. Darwin's theory of female choice as
incompetent to account for the brilliant colours and markings
of the higher animals, the preponderance of these colours and
markings in the male sex, and their display during periods
of activity or excitement, I may be asked what explanation
I have to offer as a preferable substitute.   In my *Tropical
Nature* I have already indicated such a theory, which I will
now briefly explain, supporting it by some additional facts
and arguments, which appear to me to have great weight, and
for which I am mainly indebted to a most interesting and
suggestive posthumous work by Mr. Alfred Tylor.[2]

The fundamental or ground colours of animals are, as has
been shown in preceding chapters, very largely protective,
and it is not improbable that the primitive colours of all
animals were so.   During the long course of animal development other modes of protection than concealment by harmony
of colour arose, and thenceforth the normal development of
colour due to the complex chemical and structural changes
ever going on in the organism, had full play; and the colours
thus produced were again and again modified by natural selection
for purposes of warning, recognition, mimicry, or special protection, as has been already fully explained in the preceding
chapters.

Mr. Tylor has, however, called attention to an important
principle which underlies the various patterns or ornamental
markings of animals—namely, that diversified coloration
follows the chief lines of structure, and changes at points, such
as the joints, where function changes.   He says, "If we
take highly decorated species—that is, animals marked by

---

[1] *Descent of Man*, pp. 401, 402.
[2] *Coloration in Animals and Plants*, London, 1886.

alternate dark or light bands or spots, such as the zebra, some deer, or the carnivora, we find, first, that the region of the spinal column is marked by a dark stripe ; secondly, that the regions of the appendages, or limbs, are differently marked ; thirdly, that the flanks are striped or spotted, along or between the regions of the lines of the ribs ; fourthly, that the shoulder and hip regions are marked by curved lines ; fifthly, that the pattern changes, and the direction of the lines, or spots, at the head, neck, and every joint of the limbs ; and lastly, that the tips of the ears, nose, tail, and feet, and the eye are emphasised in colour.   In spotted animals the greatest length of the spot is generally in the direction of the largest development of the skeleton."

This structural decoration is well seen in many insects.   In caterpillars, similar spots and markings are repeated in each segment, except where modified for some form of protection. In butterflies, the spots and bands usually have reference to the form of the wing and the arrangement of the nervures ; and there is much evidence to show that the primitive markings are always spots in the cells, or between the nervures, or at the junctions of nervures, the extension and coalescence of these spots forming borders, bands, or blotches, which have become modified in infinitely varied ways for protection, warning, or recognition.   Even in birds, the distribution of colours and markings follows generally the same law.   The crown of the head, the throat, the ear-coverts, and the eyes have usually distinct tints in all highly coloured birds ; the region of the furcula has often a distinct patch of colour, as have the pectoral muscles, the uropygium or root of the tail, and the under tail-coverts.[1]

Mr. Tylor was of opinion that the primitive form of ornamentation consisted of spots, the confluence of these in certain directions forming lines or bands; and, these again, sometimes coalescing into blotches, or into more or less uniform tints covering a large portion of the surface of the body.   The young lion and tiger are both spotted ; and in the Java hog (Sus vittatus) very young animals are banded, but have spots over the shoulders and thighs.   These spots run into stripes

[1] *Coloration of Animals*, Pl. X, p. 90 ; and Pls. II, III, and IV, pp. 30, 40, 42.

as the animal grows older; then the stripes expand, and
at last, meeting together, the adult animal becomes of a
uniform dark brown colour. So many of the species of
deer are spotted when young, that Darwin concludes the
ancestral form, from which all deer are derived, must have
been spotted. Pigs and tapirs are banded or spotted when
young; an imported young specimen of Tapirus Bairdi
was covered with white spots in longitudinal rows, here
and there forming short stripes.[1] Even the horse, which
Darwin supposes to be descended from a striped animal,
is often spotted, as in dappled horses; and great numbers
show a tendency to spottiness, especially on the haunches.

Ocelli may also be developed from spots, or from bars, as
pointed out by Mr. Darwin. Spots are an ordinary form of
marking in disease, and these spots sometimes run together,
forming blotches. There is evidence that colour markings are
in some way dependent on nerve distribution. In the disease
known as frontal herpes, an eruption occurs which corresponds
exactly to the distribution of the ophthalmic division of the
fifth cranial nerve, mapping out all its little branches even
to the one which goes to the tip of the nose. In a Hindoo
suffering from herpes the pigment was destroyed in the arm
along the course of the ulnar nerve, with its branches along
both sides of one finger and the half of another. In the leg
the sciatic and scaphenous nerves were partly mapped out,
giving to the patient the appearance of an anatomical
diagram.[2]

These facts are very interesting, because they help to
explain the general dependence of marking on structure which
has been already pointed out. For, as the nerves everywhere
follow the muscles, and these are attached to the various bones,
we see how it happens, that the tracts in which distinct
developments of colour appear, should so often be marked out
by the chief divisions of the bony structure in vertebrates, and
by the segments in the annulosa. There is, however, another
correspondence of even greater interest and importance.
Brilliant colours usually appear just in proportion to the

---

[1] See coloured Fig. in *Proc. Zool. Soc.*, 1871, p. 626.
[2] A. Tylor's *Coloration*, p. 40; and Photograph in Hutchinson's *Illustrations of Clinical Surgery*, quoted by Tylor.

development of tegumentary appendages.   Among birds the
most brilliant colours are possessed by those which have
developed frills, crests, and elongated tails like the humming-
birds; immense tail-coverts like the peacock; enormously
expanded wing-feathers, as in the argus-pheasant; or magnificent plumes from the region of the coracoids in many of
the birds of paradise.   It is to be noted, also, that all these
accessory plumes spring from parts of the body which, in
other species, are distinguished by patches of colour; so that
we may probably impute the development of colour and of
accessory plumage to the same fundamental cause.

Among insects, the most brilliant and varied coloration
occurs in the butterflies and moths, groups in which the wing-
membranes have received their greatest expansion, and whose
specialisation has been carried furthest in the marvellous scaly
covering which is the seat of the colour.   It is suggestive, that
the only other group in which functional wings are much
coloured is that of the dragonflies, where the membrane is
exceedingly expanded.   In like manner, the colours of beetles,
though greatly inferior to those of the lepidoptera, occur in a
group in which the anterior pair of wings has been thickened
and modified in order to protect the vital parts, and in which
these wing-covers (elytra), in the course of development in the
different groups, must have undergone great changes, and have
been the seat of very active growth.

The Origin of Accessory Plumes.

Mr. Darwin supposes, that these have in almost every case
been developed by the preference of female birds for such
males as possessed them in a higher degree than others; but
this theory does not account for the fact that these plumes
usually appear in a few definite parts of the body.   We
require some cause to initiate the development in one part
rather than in another.   Now, the view that colour has arisen
over surfaces where muscular and nervous development is
considerable, and the fact that it appears especially upon the
accessory or highly developed plumes, leads us to inquire whether
the same cause has not primarily determined the development
of these plumes.   The immense tuft of golden plumage in the
best known birds of paradise (Paradisea apoda and P. minor)

springs from a very small area on the side of the breast.   Mr.
Frank E. Beddard, who has kindly examined a specimen for
me, says that "this area lies upon the pectoral muscles, and
near to the point where the fibres of the muscle converge
towards their attachment to the humerus.   The plumes arise,
therefore, close to the most powerful muscle of the body, and
near to where the activities of that muscle would be at a
maximum.   Furthermore, the area of attachment of the plumes
is just above the point where the arteries and nerves for the
supply of the pectoral muscles, and neighbouring regions,
leave the interior of the body.   The area of attachment of
the plume is, also, as you say in your letter, just above the
junction of the coracoid and sternum."   Ornamental plumes
of considerable size rise from the same part in many other
species of paradise birds, sometimes extending laterally in front,
so as to form breast shields.   They also occur in many humming-
birds, and in some sun-birds and honey-suckers; and in all these
cases there is a wonderful amount of activity and rapid movement, indicating a surplus of vitality, which is able to manifest
itself in the development of these accessory plumes.[1]

In a quite distinct set of birds, the gallinaceæ, we find the
ornamental plumage usually arising from very different parts, in
the form of elongated tail-feathers or tail-coverts, and of ruffs
or hackles from the neck.   Here the wings are comparatively
little used, the most constant activities depending on the legs,
since the gallinaceæ are pre-eminently walking, running, and
scratching birds.   Now the magnificent train of the peacock
—the grandest development of accessory plumes in this order
—springs from an oval or circular area, about three inches in
diameter, just above the base of the tail, and, therefore,
situated over the lower part of the spinal column near the
insertion of the powerful muscles which move the hind limbs
and elevate the tail.   The very frequent presence of neck-ruffs
or breast-shields in the males of birds with accessory plumes
may be partly due to selection, because they must serve as a
protection in their mutual combats, just as does the lion's or the
horse's mane.   The enormously lengthened plumes of the bird
of paradise and of the peacock can, however, have no such use,

[1] For activity and pugnacity of humming-birds, see *Tropical Nature*, pp.
130, 213.

but must be rather injurious than beneficial in the bird's ordinary life. The fact that they have been developed to so great an extent in a few species is an indication of such perfect adaptation to the conditions of existence, such complete success in the battle for life, that there is, in the adult male at all events, a surplus of strength, vitality, and growth-power which is able to expend itself in this way without injury. That such is the case is shown by the great abundance of most of the species which possess these wonderful superfluities of plumage. Birds of paradise are among the commonest birds in New Guinea, and their loud voices can be often heard when the birds themselves are invisible in the depths of the forest; while Indian sportsmen have described the peafowl as being so abundant, that from twelve to fifteen hundred have been seen within an hour at one spot; and they range over the whole country from the Himalayas to Ceylon. Why, in allied species, the development of accessory plumes has taken different forms, we are unable to say, except that it may be due to that individual variability which has served as the starting-point for so much of what seems to us strange in form, or fantastic in colour, both in the animal and vegetable world.

Development of Accessory Plumes and their Display.

If we have found a *vera causa* for the origin of ornamental appendages of birds and other animals in a surplus of vital energy, leading to abnormal growths in those parts of the integument where muscular and nervous action are greatest, the continuous development of these appendages will result from the ordinary action of natural selection in preserving the most healthy and vigorous individuals, and the still further selective agency of sexual struggle in giving to the very strongest and most energetic the parentage of the next generation. And, as all the evidence goes to show that, so far as female birds exercise any choice, it is of "the most vigorous, defiant, and mettlesome male," this form of sexual selection will act in the same direction, and help to carry on the process of plume development to its culmination. That culmination will be reached when the excessive length or abundance of the plumes begins to be injurious to the bearer of them; and it may be this check to the further lengthening of the peacock's train that has led to the broadening of the feathers at the ends, and the consequent production of the magnificent eye-spots which now form its crowning ornament.

The display of these plumes will result from the same causes which led to their production.   Just in proportion as the feathers themselves increased in length and abundance, the skin-muscles which serve to elevate them would increase also ; and the nervous development as well as the supply of blood to these parts being at a maximum, the erection of the plumes would become a habit at all periods of nervous or sexual excitement.   The display of the plumes, like the existence of the plumes themselves, would be the chief external indication of the maturity and vigour of the male, and would, therefore, be necessarily attractive to the female.   We have, thus, no reason for imputing to her any of those æsthetic emotions which are excited in us, by the beauty of form, colour, and pattern of these plumes ; or the still more improbable æsthetic tastes, which would cause her to choose her mate on account of minute differences in their forms, colours, or patterns.

As co-operating causes in the production of accessory ornamental plumes, I have elsewhere suggested [1] that crests and other erectile feathers may have been useful in making the bird more formidable in appearance, and thus serving to frighten away enemies ; while long tail or wing feathers might serve to distract the aim of a bird of prey.   But though this might be of some use in the earlier stages of their development, it is probably of little importance compared with the vigour and pugnacity of which the plumes are the indication, and which enable most of their possessors to defend themselves against the enemies which are dangerous to weaker and more timid birds.   Even the tiny humming-birds are said to attack birds of prey that approach too near to their nests.

The Effect of Female Preference will be Neutralised by Natural Selection.

The various facts and arguments now briefly set forth, afford an explanation of the phenomena of male ornament,

[1] *Tropical Nature*, p. 209.   In Chapter V of this work the views here advocated were first set forth, and the reader is referred there for further details.

336

as being due to the general laws of growth and development, and make it unnecessary to call to our aid so hypothetical a cause as the cumulative action of female preference. There remains, however, a general argument, arising from the action of natural selection itself, which renders it almost inconceivable that female preference could have been effective in the way suggested; while the same argument strongly supports the view here set forth. Natural selection, as we have seen in our earlier chapters, acts perpetually and on an enormous scale in weeding out the "unfit" at every stage of existence, and preserving only those which are in all respects the very best. Each year, only a small percentage of young birds survive to take the place of the old birds which die; and the survivors will be those which are best able to maintain existence from the egg onwards, an important factor being that their parents should be well able to feed and protect them, while they themselves must in turn be equally able to feed and protect their own offspring. Now this extremely rigid action of natural selection must render any attempt to select mere ornament utterly nugatory, unless the most ornamented always coincide with "the fittest" in every other respect; while, if they do so coincide, then any selection of ornament is altogether superfluous. If the most brightly coloured and fullest plumaged males are *not* the most healthy and vigorous, have *not* the best instincts for the proper construction and concealment of the nest, and for the care and protection of the young, they are certainly not the fittest, and will not survive, or be the parents of survivors. If, on the other hand, there *is* generally this correlation—if, as has been here argued, ornament is the natural product and direct outcome of superabundant health and vigour, then no other mode of selection is needed to account for the presence of such ornament. The action of natural selection does not indeed disprove the existence of female selection of ornament as ornament, but it renders it entirely ineffective; and as the direct evidence for any such female selection is almost *nil*, while the objections to it are certainly weighty, there can be no longer any reason for upholding a theory which was provisionally useful in calling attention to a most curious and suggestive body of facts, but which is now no longer tenable.

The term "sexual selection" must, therefore, be restricted to the direct results of male struggle and combat. This is really a form of natural selection, and is a matter of direct observation; while its results are as clearly deducible as those of any of the other modes in which selection acts. And if this restriction of the term is needful in the case of the higher animals it is much more so with the lower. In butterflies the weeding out by natural selection takes place to an enormous extent in the egg, larva, and pupa states; and perhaps not more than one in a hundred of the eggs laid produces a perfect insect which lives to breed. Here, then, the impotence of female selection, if it exist, must be complete; for, unless the most brilliantly coloured males are those which produce the best protected eggs, larvæ, and pupæ, and unless the particular eggs, larvæ, and pupæ, which are able to survive, are those which produce the most brilliantly coloured butterflies, any choice the female might make must be completely swamped. If, on the other hand, there *is* this correlation between colour development and perfect adaptation to conditions in all stages, then this development will necessarily proceed by the agency of natural selection and the general laws which determine the production of colour and of ornamental appendages.[1]

General Laws of Animal Coloration.

The condensed account which has now been given of the phenomena of colour in the animal world will sufficiently show the wonderful complexity and extreme interest of the subject; while it affords an admirable illustration of the importance of the great principle of utility, and of the effect of the theories of natural selection and development in giving a new interest

[1] The Rev. O. Pickard-Cambridge, who has devoted himself to the study of spiders, has kindly sent me the following extract from a letter, written in 1869, in which he states his views on this question:—

"I myself doubt that particular application of the Darwinian theory which attributes male peculiarities of form, structure, colour, and ornament to female appetency or predilection. There is, it seems to me, undoubtedly something in the male organisation of a special, and sexual nature, which, of its own vital force, develops the remarkable male peculiarities so commonly seen, and of no imaginable use to that sex. In as far as these peculiarities show a great vital power, they point out to us the finest and strongest individuals of the sex, and show us which of them would most certainly appropriate to themselves the best and greatest number of females, and leave behind them the strongest and greatest number of

to the most familiar facts of nature.   Much yet remains to be done, both in the observation of new facts as to the relations between the colours of animals and their habits or economy, and, more especially, in the elucidation of the laws of growth which determine changes of colour in the various groups; but so much is already known that we are able, with some confidence, to formulate the general principles which have brought about all the beauty and variety of colour which everywhere delight us in our contemplation of animated nature.   A brief statement of these principles will fitly conclude our exposition of the subject.

1. Colour may be looked upon as a necessary result of the highly complex chemical constitution of animal tissues and fluids.   The blood, the bile, the bones, the fat, and other tissues have characteristic, and often brilliant colours, which we cannot suppose to have been determined for any special purpose, as colours, since they are usually concealed.   The external organs, with their various appendages and integuments, would, by the same general laws, naturally give rise to a greater variety of colour.

2. We find it to be the fact that colour increases in variety and intensity as external structures and dermal appendages become more differentiated and developed.   It is on scales, hair, and especially on the more highly specialised feathers, that colour is most varied and beautiful; while among insects colour is most fully developed in those whose wing membranes are most expanded, and, as in the lepidoptera, are clothed with highly specialised scales.   Here, too, we find an additional mode of colour production in transparent lamellæ or in fine surface striæ which, by the laws of interference, produce the wonderful metallic hues of so many birds and insects.

progeny.   And here would come in, as it appears to me, the proper application of Darwin's theory of Natural Selection; for the possessors of greatest vital power being those most frequently produced and reproduced, the external signs of it would go on developing in an ever-increasing exaggeration, only to be checked where it became really detrimental in some respect or other to the individual."

This passage, giving the independent views of a close observer—one, moreover, who has studied the species of an extensive group of animals both in the field and in the laboratory—very nearly accords with my own conclusions above given; and, so far as the matured opinions of a competent naturalist have any weight, afford them an important support.

3. There are indications of a progressive change of colour, perhaps in some definite order, accompanying the development of tissues or appendages. Thus spots spread and fuse into bands, and when a lateral or centrifugal expansion has occurred—as in the termination of the peacocks' train feathers, the outer web of the secondary quills of the Argus pheasant, or the broad and rounded wings of many butterflies—into variously shaded or coloured ocelli. The fact that we find gradations of colour in many of the more extensive groups, from comparatively dull or simple to brilliant and varied hues, is an indication of some such law of development, due probably to progressive local segregation in the tissues of identical chemical or organic molecules, and dependent on laws of growth yet to be investigated.

4. The colours thus produced, and subject to much individual variation, have been modified in innumerable ways for the benefit of each species. The most general modification has been in such directions as to favour concealment when at rest in the usual surroundings of the species, sometimes carried on by successive steps till it has resulted in the most minute imitation of some inanimate object or exact mimicry of some other animal. In other cases bright colours or striking contrasts have been preserved, to serve as a warning of inedibility or of dangerous powers of attack. Most frequent of all has been the specialisation of each distinct form by some tint or marking for purposes of easy recognition, especially in the case of gregarious animals whose safety largely depends upon association and mutual defence.

5. As a general rule the colours of the two sexes are alike; but in the higher animals there appears a tendency to deeper or more intense colouring in the male, due probably to his greater vigour and excitability. In many groups in which this superabundant vitality is at a maximum, the development of dermal appendages and brilliant colours has gone on increasing till it has resulted in a great diversity between the sexes; and in most of these cases there is evidence to show that natural selection has caused the female to retain the primitive and more sober colours of the group for purposes of protection.

[*Editor's Note:* The Concluding Remarks have been omitted.]

23

Reprinted from pages 11–13, 14–16, and 24–43 of *Sexual Dimorphism in the Animal Kingdom: A Theory of the Evolution of Secondary Sexual Characters,* Black, London, 1900

INTRODUCTION

J. T. Cunningham

[*Editor's Note:* In the original, material precedes and follows these excerpts.]

It may be truly said that no animal is without adaptations; it must be provided with some means by which the essentials of life are secured, but these means may be exceedingly simple or exceedingly complex. But there is another idea implied in the conception of adaptation, the idea of unity in diversity, of parts essentially similar being modified in different animals for different purposes, being adapted in many cases for purposes quite different from that which they originally served, as in the case of the fore-leg becoming in birds the wing. It is possible to trace such modifications and more or less disguised homologies without any reference to the doctrine of evolution. The principle of descent with modification gives the explanation of the phenomenon. The unity of plan is due to heredity, the divergence of detail to adaptation under different conditions. But we must pursue the investigation further and endeavour to discover how the modification is effected. It may be asked, since we admit that adaptation is such a prevalent phenomenon in the animal kingdom, even if it is not universal and exclusive, what other explanation of it is required than natural selection ? In reply to this I would urge in the first place that natural selection implies and assumes the appearance of variations, of slight modifications, by virtue of which certain individuals differ from their brethren and from their parents, in fact from all pre-existing individuals. The real explanation of evolution therefore lies in the explanation of individual variations. We admit that they occur, but how and why ?

Darwin held that the use or disuse of organs and the direct action of conditions caused modifications of individuals in definite directions, and that these modifications were hereditary in some degree. Now, if once we admit this, selection becomes a secondary and subordinate factor. For if a new set of conditions, or a change of habits, caused a hereditary change of structure in all the individuals exposed to it, continuous modification would take place even if all the individuals generated survived, or if those which were killed were taken at random without selection. A later school of evolutionists have maintained that the effects of habits or conditions on the individual are not inherited, and therefore not cumulative. According to this view only those variations are hereditary which arise in the germ, in the internal constitution of the egg; such variations are supposed to be numerous and to take place in all possible directions, and natural selection is supposed to pick out from among them those which are advantageous, and so accumulate them. I do not propose here to discuss the various theories of heredity. The question of the possibility of the transmission of acquired characters, of the determination of congenital modifications by the direct influence of conditions, is a very important one, and has been much discussed. But I wish to draw attention to a mode of considering the subject which is generally neglected, namely the inductive method. The doctrine of evolution is an induction from the facts of zoology; in my opinion conclusions concerning the method and the causes of evolution can also be obtained as inductions from a sufficiently wide survey of the phenomena.

Every one will admit without hesitation that all variations must be due to causes. But according to the selectionists the conditions of life are never the causes of hereditary variations. On this view, therefore, no modification which is hereditary can have any essential or necessary relation to the requirements of life. The fitness of a structure or structural mechanism for its function is thus originally accidental. For example, the long legs of wading birds are suitable for their habit of wading, but the habit of wading did not make the legs long. Some individuals had longer legs than others, and those which had the longest survived because the length of their legs enabled them to wade.

Romanes maintained with much truth that natural selection was a theory of the origin of adaptations, not necessarily of the origin of species, but it is further necessary to realise that it originates adaptations only in the sense of preserving and combining the variations or modifications which occur, and which happen to be advantageous. It may be said to combine only in the sense of causing different variations in the parents to be transmitted together to the offspring, and of allowing new variations to occur only in the individuals which have survived.

[*Editor's Note:* Material has been omitted at this point.]

A special adaptation necessarily implies special habits and special stimulations. These must produce, at least in the individual, corresponding modifications of structure. The peculiarities of special adaptations are generally of the same kind as the modifications which would result from the direct action of the special conditions. The natural conclusion is that the peculiarities in question are the effects of the special conditions accumulated by heredity. The only reason for rejecting this conclusion is the belief that the direct effects of conditions are not inherited. But this is only a belief, not an established truth.

Now it is possible by actual observation to ascertain what evidence there is that variations which might by natural selection be combined into the adaptations we see, do occur apart from the special habits or conditions to which the adaptations are related. The variations that occur constantly in the form of individual differences have been minutely investigated in the past few years by statistical methods with the aid of the higher mathematics. The greater the difference the more rarely it occurs, and occasionally striking abnormalities are observed, the character of which points to definite principles of symmetry and repetition in development. But it is not proved that, without change of conditions, variations occur which could, by selection, give rise to such special adaptations as abound in the animal kingdom. For example the power of partial or complete flight by means of a membranous fold of skin has been evolved in many independent cases in the vertebrate sub-kingdom, in the extinct pterodactyle reptiles, in bats, in flying squirrels, and in flying marsupials. But the variations in the condition of the skin, in animals that do not fly or take long leaps through the air, are not such as to justify the belief that these variations would make any difference in the struggle for existence when long leaps became necessary. Unless the differences were great enough to cause the comparative success of certain individuals and the comparative failure of others, the process of natural selection could not commence. On the other hand, the very condition which the principle of natural selection in this case implies, the necessity or advantage of long leaps or incipient flight, involves the practice of new and special actions and movements. Such new habits must have had some effect in modifying the muscles, bones, and skin, and on the *hypothesis* that these effects in time became hereditary the evolution becomes intelligible. Thus it seems to me much more probable that the new conditions produced the successive steps of modification directly, than that they merely selected variations already in existence.

[*Editor's Note:* Material has been omitted at this point.]

I may possibly in the future be able to undertake a more detailed study of the evidence concerning the action of the external conditions of life in producing the differences of structure in different kinships of animals, and in different stages of the individual life. But the present work is devoted especially to the consideration of the commonest case of Polymorphism, namely, Sexual Dimorphism, and the general aspects of this subject are to be exhibited in the remainder of this Introduction.

Darwin's theory of Sexual Selection is a logical corollary of his more general theory of Natural Selection. The facts on which it is based cannot be denied. Fierce combats between mature male animals for the possession of the females are known to occur in large numbers of species. The average result of such struggles must be the victory of those individuals best provided with weapons, and with the strength and skill to use them. Rivalry of another kind in the endeavour to please the female and win her admiration and consent, is displayed abundantly in the higher vertebrates generally.

Special decorations or peculiarities in the external appearance or appendages of the body are seen to have no significance except in courtship, and it is not unreasonable to infer that the individuals best endowed with beauty or melodiousness of voice succeed oftenest, and impress their qualities on the next generation more than their less fortunate fellows. Assuming, then, that differences in the degree of development of such peculiarities occur among the males in each generation, and that the best succeed in producing most offspring, transmitting their advantages in various degrees to their male progeny, the progressive evolution of such characters is the necessary consequence.

The theory, however, has not been universally accepted by followers of Darwin. Mr. A. R. Wallace has been a vigorous and consistent opponent of it. His objections are clearly summarised and criticised by Romanes in his *Darwin and after Darwin,* vol. i. 1892. One of them is, that the supposed action of sexual selection would, if it occurred, be wholly neutralised by the action of natural selection; for unless the most highly-ornamented males preferred by the females are also the most fitted for the other conditions of life, they will be eliminated in the general struggle for existence, and the chances must be small that the otherwise best fitted males should be likewise the most highly-ornamented, unless there is a natural correlation between embellishment and general perfection, which Mr. Wallace maintains to be the case. Now this objection is not a very powerful one, for it was obvious from the first that the female could only exercise her choice among the individuals which had survived in the struggle for existence, and Darwin's theory simply maintains that after a male has succeeded in getting a living and escaping his enemies, he must compete for the possession of a mate before he can leave progeny.

Another objection of Mr. Wallace's is that each bird finds a mate under all circumstances; in other words, that failure to please one female does not imply failure to please all. But, nevertheless, it would seem probable, especially in polygamous animals, that some males are fathers on a much larger scale than others.

Again, he considers it very improbable that through thousands of generations all the females of a species should always have had the same taste with regard to every detail in the nuptial adornment of their mates. This is certainly a valid objection, or at least appears to be so. The criticism of Mr. Romanes is that it is not very safe to infer what sentiments may be in the mind of a hen. We have evidence that the preferences of females with regard to male attractions are very conservative in our own species, but it is certainly difficult to conceive that the female Argus pheasant should have the power or the opportunity of appreciating minute differences in the perfection of the ocelli on the wing-feathers of the male.

Mr. Wallace argues that the principal cause of the greater brilliancy of male animals in general, and of male birds in particular, is, that they do not stand so much in need of protection by concealment as the females do. Therefore natural selection is more active in repressing in the female those bright colours which are normally produced in both sexes by general laws. More than this, natural selection promotes the development of heightened colour in the males, because this is correlated with health and vigour, and the healthy and vigorous males are preserved by natural selection.

With regard to the display which, according to Mr. Romanes, is the strongest of all Mr. Darwin's arguments in favour of sexual selection, Mr. Wallace says that there is no evidence that the females are in any way affected by it. On these points I cannot agree with either Romanes or Wallace. I do not consider that the display is the strongest argument in favour of Darwin's theory, and I think the effect on the females is perfectly obvious. Wallace argues that the display and gestures of the male, in the case of birds, for example, may be due to general excitement, pointing out that moveable feathers are habitually erected under the influence of anger and rivalry to make the bird look more formidable in the eyes of antagonists. The display and erection of organs or appendages present only in the males according to this argument is no proof of competition or of selection on the part of the females, no proof that the most adorned males accomplish the act of reproduction while the least ornamented fail to beget offspring.

The more brilliant colour and excessive development of certain feathers or other appendages, being due according to Wallace to the exuberant health and vigour of the males, the question arises how is it that the more vigorous development does not extend equally to all parts and appendages of the body, instead of being confined, as it is very precisely, to particular parts or appendages ? To explain this fact Wallace adopts the theory expounded by Mr. Alfred Tyler in his book on *Coloration of Animals and Plants* (1886). Mr. Tyler maintained that diversified coloration follows the chief lines of structure and changes at points, such as the joints, where function changes. Mr. Wallace maintains that the enlarged plumes of male birds are situated at points where the nerves and blood-vessels converge. Thus, in his book entitled *Darwinism* he states that Mr. Beddard examined for him the anatomical relations of the roots of the great tufts of golden feathers in the bird of paradise, and found that they were situated just above the point where the arteries and nerves for the supply of the pectoral muscles leave the interior of the body. But, as Professor Lloyd

Morgan has already pointed out, there is no physiological reason why the growth of feathers should be connected with the proximity of main arteries and nerves. We have no reason to suppose that the large supply of blood for the pectoral muscles overflows into the skin and causes the feathers of the breast to sprout luxuriantly as does the grass round a spring of water. And, moreover, if it were so, the special plumes ought to be equally developed in the females, and indeed all birds ought to possess plumes on the breast similar to those of the male bird of paradise, since in all the position of the nerves and blood-vessels is the same.

In so far as it attributes the evolution of secondary sexual peculiarities to supposed general physiological principles, and not to selection from indefinite individual variations, Wallace's explanation resembles these views of evolution which are distinguished as Lamarckian. This is only one out of several subjects on which Mr. Wallace is Lamarckian without knowing it, and apparently in spite of himself. Although he independently invented the theory of natural selection he does not admit the action of that principle in the case of the mental and moral characteristics of man, or of a special kind of selection in relation to sex. Not that Mr. Wallace has ever acknowledged himself to be a Lamarckian. He probably has a strong prejudice against the epithet, but that may be because, like many others, he has to some extent misunderstood the position of the Lamarckians, and has not realised the similarity between his own views and theirs.

The views which I am about to support are also of the kind called Lamarckian, but they are by no means the same as those of Mr. Wallace on the subject under consideration. I attribute the evolution of secondary sexual characters to alleged physiological principles of growth and modification, but these principles are quite different from those suggested by Mr. Wallace, and I believe they are principles founded on sufficient evidence, whereas, after careful consideration, I can find no evidence of any necessary or general connection between health and vigour, or anatomical features, and the special peculiarities in which one sex differs from the other.

Other suggestions or theories which have been propounded in explanation of the evolution of secondary sexual characters appear to be merely ingenious variations of the idea of selection. Stolzmann[1] has maintained that among birds the males are more numerous than the females, which of course is one of the main points in Darwin's theory. According to Stolzmann the fact is a necessary consequence of the defective or scanty nourishment of the eggs. The female devoting her energies and labour to the work of nest-building takes less food and performs more work than she otherwise would, and consequently the nutrition of the eggs is less generous. Scantily-nourished eggs produce males rather than females, and the majority of birds' eggs being scantily nourished, males are in a majority among the young produced. It is obvious that this argument does not deserve serious attention. It is certain that birds' eggs, as compared for instance with those of fishes, amphibia, or mammals, are supplied with an enormous quantity of the most nutritious food-yolk, and it is difficult under these circumstances to understand what is meant by the assertion that they are badly nourished.

However, to continue Stolzmann's argument, the excess of male birds is pernicious to the species, not only on account of their interference with the females during the performance of their maternal duties, but also because they occupy valuable space and lessen the supplies of food. Anything therefore tending to lessen the undue excess of males would be seized upon and perpetuated by natural selection. Hence the gaudy colours, crests and spurs, and pugnacious habits of the males. The bright colours render them visible not

[1] *Proc. Zool. Soc.* 1885.

only to each other but to hawks and other enemies; their long plumes diminish their powers of escape; their constant fights tend to diminish their numbers, while their dances and rival singing are merely distractions which protect the females against their too constant attentions.

A very little reflection is enough to show that these curious and ingenious suggestions are of no real assistance towards the solution of the problem. As far as individual selection within the species is concerned, the males with the special peculiarities most highly developed must according to the theory be the first to be exterminated, and therefore it is impossible to understand how such peculiarities could ever make much progress in evolution. The theory is founded on the argument that the decimation of the males is for the good of the species, which merely means that those species in which the males are most conspicuous, and exposed to the most numerous perils, will flourish most. But how are we to conceive of the evolution of a species which is at the same time characterised by the greatest development of peculiarities in the males and the greatest destruction of males on account of those very peculiarities. The two processes supposed are in direct opposition, and logically the result would be nothing. There is no variation of species apart from individuals, and according to the assumptions made by Stolzmann, the individuals in the species with the greatest development of the required characters would not transmit their peculiarities.

The fundamental objection to Darwin's theory of sexual selection, or to any other selection theory, is that it does not account for the origin of the variations which it assumes. Those who consider the process of selection as the most important factor, maintain, and rightly maintain, that individual variations do actually occur, and they are either not inclined to analyse the problem further, or they support some theory which professes to explain the indefinite variation, or variability, on which selection acts. Now there are two almost universal peculiarities of secondary sexual characters on which the theory of sexual selection throws no light whatever: (1) the characters do not begin to appear in the individual until it is nearly adult and sexually mature, in other words they appear when, or a little before, the animal begins to breed; (2) they are inherited only by the sex which possesses them.

If it were a mere question of the occurrence of individual variations, and a preference by the female sex, there would be no reason whatever why variations occurring at the beginning of life should not be selected. The theory merely supposes that each female has a number of candidates for her favour which are not all alike. Now let us suppose that a single cock-bird, a peacock for example, began to develop its special tail as soon as it developed its other feathers, and had a well-developed tail of the characteristic kind as soon as it was fledged. When it became sexually mature its special plumage would be perfect, while that of its fellows of the same age was only partially developed, for in the ordinary case males are mature before the development of their secondary characters is completed. According to the theory of sexual selection, then, the young male in which precocious development of the special plumage occurred would be preferred to others of the same age in which the development was later, and the result in a few generations would be that the peculiarities of the males would appear as soon as the permanent feathers appeared. The selection of course does not take place until maturity is reached and mating occurs, but the selection has nothing to do with the production of the variations, and the time at which selection takes place cannot determine the time at which the modifications occur in the individual.

Secondly, why should the individual peculiarities in the

male selected not be transmitted to his female offspring as well as to the male? It is true Darwin believes that this has occurred in some instances, but why does it not occur always? This is a phenomenon with which selection has nothing to do. Wallace maintains that the females require to have dull colours, and to be inconspicuous, in order to escape the notice of enemies when sitting on their nests. But selection, whether of dull-coloured females by the conditions of life, or of brilliant and ornamented males by female fastidiousness, only assumes, and does not explain, the inheritance of special characters by one sex and not by the other.

It is with these two special peculiarities of secondary sexual characters that my own arguments are chiefly concerned, and I will therefore consider here what attempts have hitherto been made to explain them.

It cannot be said that Darwin succeeded in explaining the phenomena. In the extent and accuracy of his knowledge, and the perfect impartiality and devotion to truth with which he collected and examined the evidence bearing upon the general problems of structure and evolution, Darwin surpassed all his predecessors and contemporaries. His detailed and careful examination of the problems of evolution in general, and of sexual dimorphism in particular, was a great and noble work. His treatises form the foundation of our present knowledge on these subjects. But we do not honour him most by maintaining his doctrines and conclusions without change. There is no finality in knowledge. It is perfectly true that some new views may be in the wrong direction, may be retrogressive, but progress, further investigation of causes, is always possible, and it is no proof of the unsoundness of new conclusions that they go beyond or differ from those of any authority, however great.

Darwin's chief conclusion concerning the peculiarities of secondary sexual characters which I have mentioned, is that there is a connection between them. He found that it was a general rule that characters were inherited in the offspring at the same age at which they appeared in the parents, and that when characters appeared in one sex late in life they were generally inherited only by the same sex. He attributed this to the fact that the sexes do not differ much in constitution until the power of reproduction is gained, while after that the differerence is great. In support of this he cites the well-known contrast between ordinary species of deer and the Rein-deer. In the latter both sexes bear antlers, although those of the female are considerably smaller, and the antlers commence to appear in both sexes within four or five weeks after birth; in other deer the male alone bears antlers, and they begin to appear at periods varying from nine months to twelve or more after birth. It must be remembered, however, that reindeer as well as other deer shed and renew their antlers every year, and even in the reindeer the full size and branching is not attained till after several years. Darwin, however, concludes that one, though not the sole, cause of characters being exclusively inherited by one sex, is their development at a late age. How far this is a sufficient explanation I shall discuss in the following pages, but it is clear without further discussion that it is not an explanation of the origin of the variations themselves. Darwin does not attribute the latter to any definite causes, merely pointing out that secondary sexual characters are highly variable, and therefore afford plenty of opportunity for selection.

Later investigators have offered various suggestions as to the origin of the variations in question, for the most part attributing them to "constitutional" differences between the sexes, and basing their views largely on the fact that, as a general rule, it is the males that have been modified, while the females exhibit the original condition. Thus the American zoologist, W. K. Brooks (1883), attempted to explain unisexual variation and heredity by a modification of Darwin's "provisional hypothesis" of gemmules. According to this hypothesis, gemmules of infinitesimal size were given off by all the organs and tissues of the body, and were accumulated in the reproductive cells. It was the property of these gemmules to develop again the organs and tissues from which they were derived in the parents. Brooks suggested that the male gemmules were more prone to variation than the female, and therefore when they developed in a male individual they gave rise to variations. The male, therefore, had a special tendency to variation, while the female kept up the general constancy of the species; thus he attempted to explain why it is that often in allied species the females are much alike, while the males are very different.

St. George Mivart regards the beauty of males and their other peculiarities as the direct expressions of an internal force.

This view of the unisexual characters as due to internal constitutional causes has been elaborated by Messrs. Patrick Geddes and Arthur Thomson.[1] They argue that, essentially and constitutionally, males incline to activity, females to passivity; that the physiological processes in the female tissues tend to the passive accumulation of nourishment and to growth, while in the male they tend to the active production of movement and energy. When there is a difference the female is in many insects and lower animals large and stationary, the male small, agile, and often short-lived. The details of secondary characters are due to the abundance of diverse excretory products produced by the more intense physiological changes in the tissues of the male. To quote the words of these authors: "So brilliancy of colour, exuberance of hair and feathers, activity of scent glands, and even the development of weapons, are not and cannot be

[1] *Evolution of Sex*, 1889.

explained by sexual selection, but in origin and continued development are outcrops of a male as opposed to a female constitution." The male element of reproduction, the spermatozoon, is an extremely minute, microscopic particle, shaped like a tadpole, and moving with great activity in a liquid medium by means of the lashings of its tail; the egg is a large mass of living substance or plasm charged with nutriment, thousands or millions of times as large as the spermatozoon, and showing no movement at all except slight internal contractions and slow changes. Geddes and Thomson regard the male animal as, so to speak, made up of spermatozoa. They consider every cell in the body of the male as partaking of the active, spendthrift character of the spermatozoon, while the female tissues and the female animal as a whole are, like the egg, inert, receptive, vegetative.

It is obvious enough that none of these suggestions are of any assistance in explaining the details of the unisexual peculiarities; they do not explain why the male constitution of the stag is exhibited in the antlers on its head, in the peacock in the feathers of its tail. With regard to size, the authors admit that the superiority of the males in many mammals and birds is an objection to their theory. They hold that the apparent exceptions are the natural result of the increased stress of external activity which is thrown upon the shoulders of the males when their mates are incapacitated by incubation or pregnancy, and they point to the strengthening influence of the combats between males, and to the large supplies of nourishment which the females give up to their offspring.

Eimer also states, as one of the general principles of the origin of variations, that where new characters appear, the males, and especially the vigorous old males, acquire them first; that the females, on the contrary, remain always at a more juvenile, lower stage, and that the males transmit these new characters to the species.　He calls this the law of male preponderance.

For my own part, I am inclined to doubt whether there is any essential or constitutional difference between the sexes with regard to the tendency to variation.　In vast numbers of species the individuals of opposite sexes are so much alike that it is difficult to distinguish them without examination of the generative organs.　Examples of this fact among mammals are furnished by the mouse, the cat, hyæna, rabbit, hare, and many others.　Among birds examples are very abundant. Darwin found that there were six classes into which birds could be divided according to the differences and similarities in the characters of the sexes.　In two of these classes the adult male resembles the adult female, examples of which condition are afforded by the robin, the kingfisher, many parrots, crows and rooks, and the hedge-warbler.　In fishes the sexes are more often alike than different, and among the lower animals special peculiarities in the males are by no means universally present.

Cases in which the exceptional characters, conspicuous or extraordinary peculiarities, are confined to the females, though less common than the opposite condition, are not wanting, as in the species of *Turnix* among birds mentioned by Darwin.

On the other hand, in either sex unisexual characters have, as a general rule, some function or importance in the special habits or conditions of life of the sex in which they occur.　The antlers of stags are certainly fighting weapons, and are habitually employed by every healthy stag in combat with his fellows.　The special plumes of male birds are habitually erected and displayed in courtship.　The melody of singing birds is intended for the ears of their mates, and the note of the male bell-bird, the call-notes of birds in general, are love-notes.　So far there can be no doubt that Darwin was perfectly right and his opponents all wrong.

The facts being so, there is rivalry, combat, and competition, and there must be selection. But the important truth, which appears to have been generally overlooked, is that in the case of each special organ its special employment subjects it to special, usually mechanical, irritation or stimulation, to which other organs of the body are not subjected. Every naturalist and every physiologist admits that in the individual any irritation or stimulation regularly repeated produces some definite physiological effect, some local and special change of tissue in the way of either growth or absorption, enlargement or decrease, or change of shape. Thus not only hypothetically at some former time, but actually at present in every individual, the unisexual organs or appendages are subjected in their functional activity to special strains, contacts, and pressures, that is, to stimulation, which must and does have some physiological effect on their development and mode of growth. This argument will, of course, be received with more or less courteous contempt, because it is held by the majority that any effects on the individual, any " acquired characters," are not inherited, and therefore have nothing to do with the evolution of characters which are constant in species, and evidently hereditary. I do not propose here to attempt to prove that acquired characters are inherited. My object is merely to point out how remarkably the multitudinous facts all agree with the hypothesis that secondary sexual characters are due to the inheritance of acquired characters.

We have seen that the attempts that have been made to explain the restriction of such characters not only to one sex, but to the period of maturity in the individual and often to one period of the year, have not been successful. The true explanation in my opinion is, *that heredity causes the development of acquired characters for the most part only in that period of life and in that class of individuals in which they were originally acquired.* Heredity, according to my conception of it, whatever its mechanism, is a tendency in the new individual to pass successively through the same stages of growth as its parent. That which is inherited is not a state, but a process. Unisexual characters consist, in general, in modifications of growth in particular parts of the body, very often in excessive growth or hypertrophy. They may, in fact, in many cases, be correctly described as excrescences, and these excrescences are as truly the result in the first instance of mechanical or other irritation, as a corn in the human epidermis. But the irritation and the consequent hypertrophy or proliferation have been in every generation inseparably associated with a certain condition of the organism, that condition, namely, in which the testes, or the ovaries in certain cases, were mature and active.

If there were no question of inheritance in the matter, and the secondary sexual character were produced *de novo* by certain stimulations applied only in one sex when the animal began to breed, then we should have no cause to wonder why such characters were confined to one sex and to one period of life in that sex. The reason would be that the stimulations only acted on the individuals of the one sex at that period. The characters would be acquired characters confined to the individuals which acquired them, and not appearing until the actions and influences to which they were due came into operation.

But it is evident enough that the development of the characters at the present stage of evolution takes place by heredity, even though the usual irritations are partly or entirely wanting. The modifications of growth which in each generation have been produced in a certain phase of individual development, occur of their own accord when that phase is reached or approached. In the more familiar cases the phase in question is the functional activity of the testes, and the physiological condition associated therewith.

The activity of the testes does not leave the rest of the body unaffected. On the contrary, it is associated with a nervous exaltation which again influences the activities of all the other organs. It was during this condition that the secondary characters were originally produced. The special processes of growth occur in subsequent generations by heredity, *when the body is in the same condition, not otherwise.* Thus we understand why it is that the antlers of the stag do not develop in the female, and fail to develop perfectly in the stag which is castrated. In either of these cases the animal fails to reach the condition of body in which alone the local hypertrophy can occur. On the other hand, it is not to be doubted that the female inherits the tendency to produce antlers. This is necessarily the case because the female individual, like the male, arises from the union of male and female germ-cells. It is also proved to be the case by the fact that the female often transmits to her progeny peculiarities in the secondary sexual characters of her own father, and even occasionally, when old and no longer fertile, develops to some extent such characters herself.

The known phenomena of the frequent development of male characters in aged females, and of the suppression of the male characters in castrated males, compelled Weismann to admit the presence of both male and female determinants, not merely in the germ, but in the somatic cells of each individual. He considered that thus under certain circumstances the latent set of determinants might become active and manifest their effects. But he does not thus surmount the difficulty presented to his theory. The question is, why should the absence of the testes or the sterility of the ovaries affect the activity of determinants in somatic cells? According to Weismann's conceptions the behaviour of the somatic cells is determined solely by the germ from which they sprang, and in that case they cannot be affected by the presence or absence of the germ-cells which are to give rise to the *following generation*. It is conceivable that the activity of one or the other set of determinants in somatic tissues should be called forth by external stimuli, such as temperature or irritation, the determinants themselves arising only in the germ. But it is not conceivable, on Weismann's hypothesis, that the removal of the testes in the young animal should affect the behaviour of the determinants for the antlers in later life, for according to the hypothesis there is no continuity between the testes and the somatic cells. And it is obvious that if the removal of the testes can affect the development of tissues in the head, the development of the latter may affect the properties of the testes. To my thinking, the suppression of male characters in consequence of castration is in itself sufficient to disprove the theory of the absolute continuity of the germ-plasm, and its absolute independence of the somatic cells. If, as Weismann supposed, the germ-cells were entirely unaffected in their essential properties by the history and circumstances of the body in which they were contained, it is impossible to conceive how the removal of the generative organs could affect the development of the tissues of that body.

Another interesting point which gives remarkable support to the view here advocated is that the existence of secondary sexual characters is, in many cases, not merely limited to the period of mature life, but actually to that part of the year in which the reproductive organs are active, that is to the breeding season. It is a familiar fact that this is the case with the antlers of the stags, which drop off in the early part of the year and are developed again in summer. The rutting season of stags occurs in autumn, and it is only during this season that they consort with the females and engage in fights with other stags. If the antlers were originally outgrowths of bone produced by the pressure of the heads of the fighting stags against each other during the rutting season, and afterwards developed constantly by the strains of the antlers of the combatants against each other, then the hereditary tendency to the periodic annual growth of the antlers would be an intelligible fact. It cannot be too strongly emphasised that hitherto no attempt has been made to give an explanation of such facts as these. The theory that I advocate does explain them, while on all other theories of evolution the occurrence of the variations is merely assumed, and no reason for the peculiar modes and times of their occurrence is even offered. I believe I may claim that my theory is, in its special details, new and original. The theory of the inheritance of acquired characters is, of course, old, and regarded by many as even extinct. But so far as I know I have not been anticipated in my elaboration of this theory into the form in which I present it, namely, *that the direct effects of regularly recurrent stimulations are sooner or later developed by heredity, but only in association with the physiological conditions under which they were originally produced*, and that this is the explanation of the limitation of particular modifications, not merely to particular species or kinships, but to particular periods in the life of the individual, to a particular sex, and even to a particular season of the year in that sex.

It may be objected to my argument that I have not proved that particular stimulations actually produce the effects I attribute to them, and still less that any such effects are ever inherited. Such an objection would show a complete misunderstanding of my object in this work and my method of reasoning. Experimental investigations of the direct effects of stimulations have been made and published, by myself among many others.[1] That stimulations do produce local

---

[1] See *Coloration of Pleuronectidæ*, Cunningham and MacMunn. *Phil. Trans.*, 1896.

modifications in the growth of living tissues is enough for my present purpose. I make no attempt in the present work to give an account of our knowledge on this subject. With regard to the question of the inheritance of such effects, it may be true that at present we have no direct or experimental proof of its occurrence, or not sufficient proof. But on the other hand its impossibility has not been established, and my argument is merely that the hypothesis of such inheritance as I have formulated it, affords the only sufficient explanation of the phenomena which I pass in review.

Weismann in one of his essays maintains that the inheritance of acquired characters being theoretically impossible, a theory of the selection of variations arising in and from the germ is our only hope of escape from a supernatural explanation of evolution. I maintain, on the contrary, that theories of selection being found on application to the facts to be insufficient for their explanation, and the theory of the inheritance of acquired characters being found to harmonise with the facts, we are logically bound to believe that such inheritance does take place, at any rate until some other explanation can be found. I do not concern myself with the question how such inheritance can be produced, *it is a fact that the modifications with which I have to deal are hereditary, and my object is to produce inductive evidence that they were determined by special stimulations.*

I proceed, therefore, to examine the most important examples of secondary sexual characters in detail, taking the chief divisions of the animal kingdom in systematic order. In this survey I shall endeavour to show that in every case the development of the special character is associated with special habits or conditions that necessarily involve stimulations closely corresponding to the character in question, and absent where the character is absent; also that there is no evidence of the origin of the character or its elements apart from the incidence of these stimulations. Doubtless there are many errors and many deficiencies in my explanations. It is a subject in which much research is required, and I should be much gratified if my suggestions tended in some degree to promote such research.

All characters or peculiarities confined to one sex only may be called unisexual characters. These may be primarily divided into three classes according to their function, their relation to the habits and conditions of life of the animals possessing them :—

(1) Weapons, organs or characters which are employed in combats with other males for the exclusive possession of the females.

(2) Allurements, organs or characters whose function is to attract the female, physiologically speaking, to excite the sexual instinct in the female. With these may be included organs or structures whose function is to seize or hold the female.

(3) Unisexual characters which are not related directly to sexual reproduction at all, but are related to the different conditions of life to which the different sexes are exposed. These may be, and perhaps are, in all cases indirectly related to sexual reproduction, since the different modes of life of the two sexes correspond to the different parts they perform in reproduction, the one sex taking care of the eggs and the other not, for instance.

In the body of the work the various classes of animals will be considered in order, and an attempt made to describe the functions performed by the unisexual characters in the life of the animal, and to ascertain the particular modes in which the habits and actions of the animal give rise to the various stimulations to which the development of the particular characters is attributed.

AUTHOR CITATION INDEX

SUBJECT INDEX

About the Editor

CARL JAY BAJEMA is professor of biology at Grand Valley State College, Allendale, Michigan. He teaches courses in human ecology, human genetics, human sexuality, evolutionary biology, and sociobiology in the Department of Biology, and the Darwinian revolution in the History of Science program. Professor Bajema earned the Ph.D. in zoology from Michigan State University in 1963.

Professor Bajema has served as Senior Population Council Fellow in Demography and Population Genetics at the University of Chicago (1966-1967); as Research Associate in Population Studies at Harvard University (1967-1972); as Research Associate in Biology at the University of California, Santa Barbara (1974); and as Visiting Professor of Anthropology at Harvard University (1974-1975).

Dr. Bajema's research interests have concentrated on the estimation of the direction and intensity of natural selection with respect to human behavior. He is editor of *Natural Selection in Human Populations* (Wiley, 1971) and coauthor with Garrett Hardin of the third edition of the introductory college biology textbook *Biology: Its Principles and Implications* (Freeman, 1978). He has edited a Benchmark Papers in Genetics volume, *Eugenics Then and Now* (Dowden, Hutchinson & Ross, 1976) and Benchmark Papers in Systematic and Evolutionary Biology volumes, *Artificial Selection and the Development of Evolutionary* Theory (Hutchinson Ross, 1982) and *Natural Selection Theory from the Speculations of the Greeks to the Quantitative Measurements of the Biometricians* (Hutchinson Ross, 1983). Professor Bajema is editing other Benchmark volumes on selection theory and is Series Editor for the Benchmark Papers in Systematic and Evolutionary Biology series.

Professor Bajema enjoys hiking, bicycling, and cross-country skiing. He resides in Grand Rapids, Michigan.